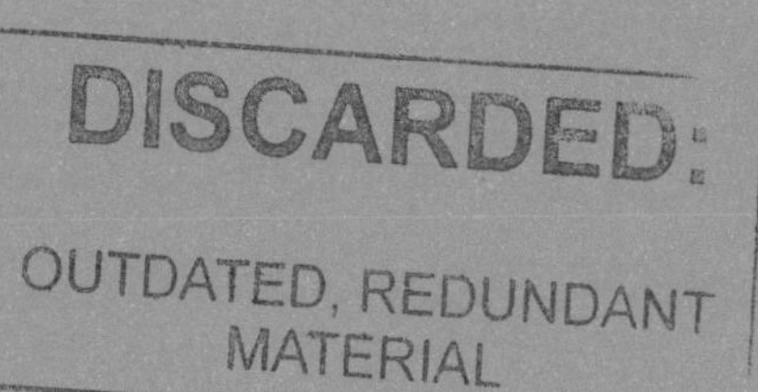
DISCARDED:
OUTDATED, REDUNDANT
MATERIAL

D1265722

FLOWER
SCHOOL

FLOWER SCHOOL

A Practical Guide to the Art of Flower Arranging

CALVERT CRARY

Executive Director of FlowerSchool NY and FlowerSchool LA

Illustrations by Monica Hellström

BLACK DOG
& LEVENTHAL
PUBLISHERS
NEW YORK

iRose

Black Dog & Leventhal Publishers
Hachette Book Group
1290 Avenue of the Americas
New York, NY 10104

www.hachettebookgroup.com
www.blackdogandleventhal.com

First Edition: November 2020

Black Dog & Leventhal Publishers is an imprint of Perseus Books, LLC, a subsidiary of Hachette Book Group, Inc. The Black Dog & Leventhal Publishers name and logo are trademarks of Hachette Book Group, Inc.

Print book interior design by Katie Benezra.

LCCN: 2020932420

ISBNs: 978-0-7624-7146-1 (paper over board); 978-0-7624-7145-4 (ebook)

Printed in China

1010

10 9 8 7 6 5 4 3 2 1

This book is for all of the wonderful supporters of FlowerSchool and is dedicated to Liz, Magalie, Marion, and Roscoe.

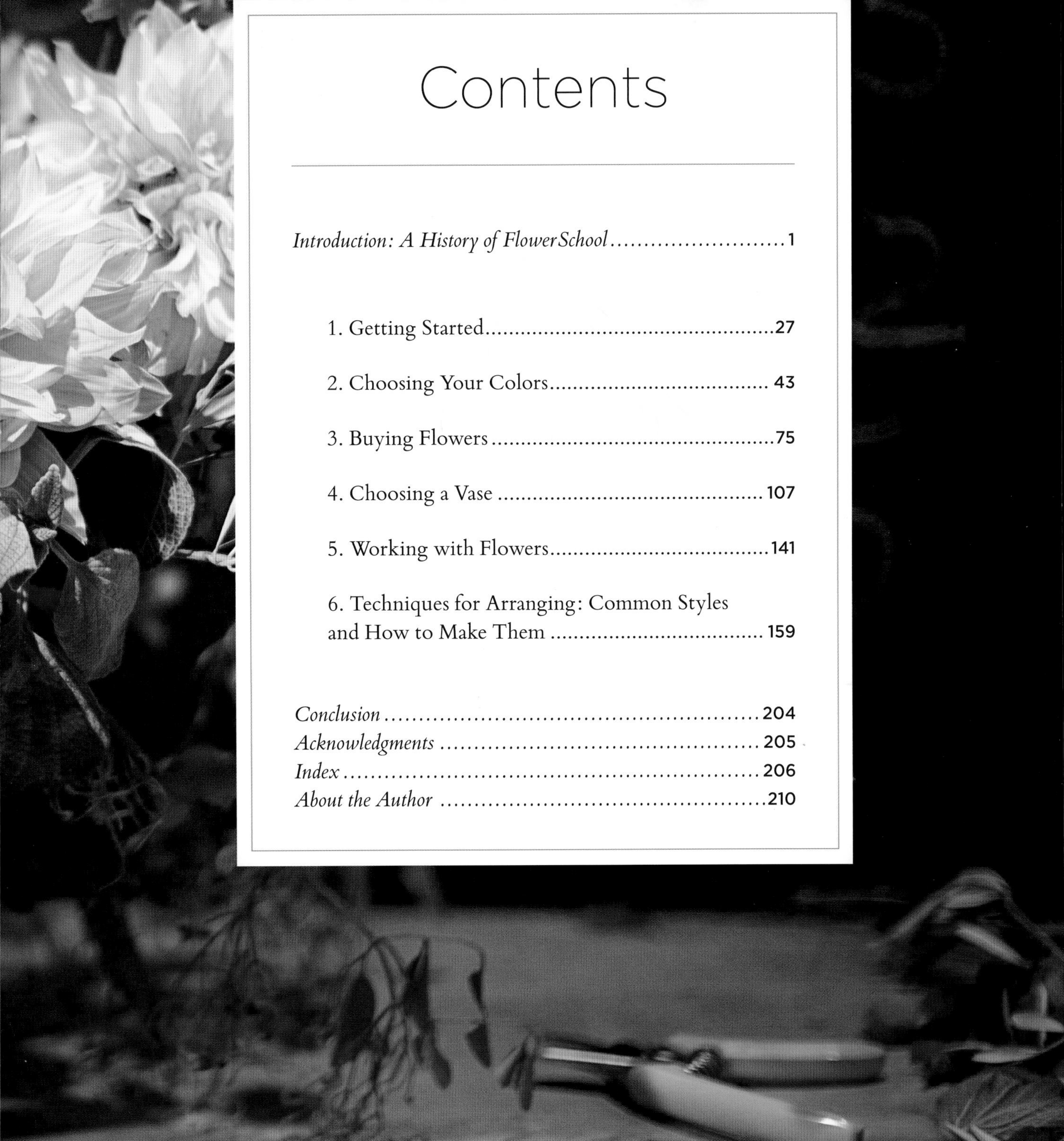

Contents

Introduction

A History of FlowerSchool

The history of FlowerSchool is similar to the history of any other trade school formed to help support an industry and celebrate its most talented practitioners. Success within a specific trade requires strong skills and practical knowledge.

The prerequisite for practicing a world-class trade is a world-class trade school. Think of George Balanchine creating the School of American Ballet in order to train dancers or world-class cooking schools in France that teach and refine young chefs. The list goes on.

The work of florists has typically been relegated to early mornings, with the goal of having their flowers "done" before any customers arrived. However, the actual process of putting these creations together can often be as wonderful as the finished product: a living work of art. The process of working with flowers often requires decision making on a moment's notice, something that can be startling for those just beginning to work with flowers. Similar to sculpting a work of art from stone, watching as each curve and angle reveals itself, a bale of branches and a bundle of flowers can turn into a masterpiece in the hands of an expert—a truly exhilarating experience when it's done right. The difference between stone and flowers is that one is fleeting and one is not.

FlowerSchool was started by Eileen Johnson in 2003 from a small flower shop in Midtown Manhattan, in a neighborhood known as Tudor City. Ms. Johnson had taken over the flower shop location from the great Michael George, one of the best florists of his time. The shop was a one-room schoolhouse with a single aim: to showcase well-known florists and connect adoring fans with their passion. As such, FlowerSchool began offering classes and started the teaching careers of great designers such as Ariella Chezar, Lewis Miller, and Remco Van Vliet to name a few.

"Make a masterpiece with a master" was the school's original tagline. Eager students were given the chance to watch as world-famous Master Florists made a bouquet, and then the students had a chance to make a similar bouquet under the master's tutelage. Because of the delicate and refined nature of this art form, the event became wildly popular—part personal and part educational. When done right, flowers can speak beyond the classroom setting. Arranging living organisms, with their own natural life force, into a pattern that resembles the spirit of the

Author Calvert Crary teaches a class at the recently opened Los Angeles outpost of FlowerSchool.

person creating the arrangement can be unbelievably invigorating. Echoing that concept, I often overhear students describe the flowers they're working with in a way that's reminiscent of how they would describe their own state of mind. "I'm putting this flower here, so it can shine or speak to me." "There you are, old friend." Wonderfully, flower arranging often includes a healthy dose of personal exploration.

A casual class or chance opportunity to cut from a garden can turn into a lifelong passion with endless possibilities. I've seen it happen. There is tremendous satisfaction to be taken from manipulating an object of beauty and helping it contribute to a higher design aesthetic. Providing an opportunity to work with such beauty is a great honor and responsibility. One that I never take lightly. Too much handling of

A FlowerSchool student makes adjustments to her arrangement during one of our weeknight classes.

flowers can destroy them, while neglecting a flower can result in an arrangement that feels unfinished and insubstantial. Working with flowers is a discipline that requires restraint as much as it requires artistic talent.

As new florists work to achieve higher levels of artistry, more questions arise. Where did these flowers come from? How were they prepped? Can I make a career of doing this? What is the technical work that goes into making flower arrangements? As more students began to ask these questions, FlowerSchool worked to develop

A group of burgeoning florists practices the fundamentals of flower arranging in the workspace at FlowerSchool's New York location.

more comprehensive programs for career development training. It's named the Floral Design Program. The program is now licensed by the Bureau of Proprietary School Supervision, a branch of the New York State Board of Education, which has made FlowerSchool an authority in floristry career education. To this day, FlowerSchool NY is the only licensed school in New York for floral design education.

Michael George was the first florist I ever saw create art using flowers. He was able to make his style come to life in unexpected ways, delivering astonishing

arrangements right before your eyes and without even a hint of hesitation. Most of the really great people I worked with in the beginning of my career came out of Michael's training program. They were all amazing technicians with a passion for perfection. As a result, the reputation of the school's training program grew and eventually entire companies were sending new staff members to begin their training at FlowerSchool. There is nothing that gives me, and everyone at FlowerSchool, a better feeling of accomplishment than to have professional companies rely on our program. Not only do students learn how to become florists, but they also learn how to do it with success.

What Makes a Master Florist

The word "master" gets thrown around with regularity regarding crafts and trades. Master carpenter, master painter, master cobbler, master colorist, and, yes, master florist. But what does it mean to be a master florist?

FlowerSchool does not take the title Master Florist lightly. A true master is a master of many aspects of doing flowers, not only the technical aspects of making a flower arrangement. FlowerSchool Master Florists are able to build inspiring concepts based on past experience, client expectations, art, culture, and essentially defining the seasons. They must be able to gather, train, and prepare the perfect team in order to execute their projects to stunning result. Master Florists are floral designers who have a unique artistic vision combined with knowledge of flowers and a deep grounding in the fundamental mechanics of working with cut flowers. Master Florists must be conversant with the life cycle of flowers, cognizant of the past history of floral art, and able to channel the future in terms of floral artistry. Master Florists can visualize a style and make a creative statement that is unique to their own particular vision.

FlowerSchool Master Florist Lewis Miller adds embellishments to an arrangement while teaching a class at FlowerSchool in New York.

One of our more advanced students at FlowerSchool in New York works on creating a large statement arrangement.

The FlowerSchool prerequisites for being a Master Florist are deceptively few but difficult to achieve. We require a minimum number of years in the business with a company that has a well-trained staff who can execute that company's vision without fail and that has a superior aesthetic vision regarding color, form, and design. As anyone who has accomplished this knows, it takes a huge amount of ambition, dedication, and passion to achieve a mastery of any craft, and flower arranging is no different.

A young woman approached the school one day asking for a teaching position at FlowerSchool. As we do with all eager industry people, our team sat down to meet with her. The conversation ranged over a wide variety of topics: inspiration, workflow, and so on. At some point, we asked how long she had been working.

"Four years," she replied. Weddings had been her main focus, so we asked how many weddings she had done flowers for in a typical year; "six to seven per year," she answered. Realistically, 24–28 weddings a Master Florist does not make! Not to say that this young woman was not competent; in fact, her work was quite beautiful. But in order to really educate others from a position of authority, a Master Florist must have worked on hundreds, if not thousands, of weddings and events, working with all types of customers. After all, would you let a beautician cut your hair after giving only 24–28 haircuts?

Another indication of mastery is having a well-trained, permanent staff, not one made up of freelancers. This indicates that the company, or designer, contributes to a technical training program for their employees, so that they understand the company's style and are able to expertly and consistently execute that style. Further, you can tell if a company's or designer's style is worthy of being taught by their success in adequately training their staff. Florists who specialize in a style, develop a workflow, and pass that on to their employees are florists who have perfected their craft. These are the artists who have the most to teach.

There are many great designers, but like a great painting, a brilliant poem, or a particularly superb slice of cake, when mastery is achieved, it stands alone. Anyone can become a florist or floral designer with a steady dose of ambition, passion, and vision. But in order to become a master, you must also be able to manipulate the flowers to create a heightened level of beauty, no matter where the flowers have come from. You may not always have the products you want. Actually, you never do. But Master Florists are able to use whatever products they have on hand to make something gorgeous.

Simply put, the whole needs to be greater than the sum of all its parts. This greater whole is achieved by using every element in an arrangement in the most creative way possible. Further, in order to be a FlowerSchool Master Florist, you must also be a superior educator who is friendly and supportive, with a commitment to teaching and mentoring.

FlowerSchool Master Florist Oscar Mora helps students learn to create arrangements of all shapes, sizes, and, in this case, heights.

Becoming a florist is now a common career path that typically begins with an apprenticeship, similar to the way that chefs often begin their careers working in low-level positions in the "back of the house" at a great restaurant. By watching a great chef cook, they learn and work their way up through the ranks to the front of the house. The world of flowers is not dissimilar. In fact, Remco Van Vliet, who creates the famous flower arrangements for the Metropolitan Museum of Art in New York City, is often asked by the museum to put the finishing touches on his arrangements in view of the public so people can see how he works. Watching the process of a master as they work is one of the most wonderful parts about being a florist, and it is an essential part of the learning process.

The apprentice relationship is at the center of what it takes to be a great florist—learning the artistry and techniques of those who have come before you. As you work your way through this book, think of yourself as an apprentice. By the time you've finished reading this book, you should have the skills required to strike out on your own and create your own designs.

What You Will Learn

FlowerSchool, including the FlowerSchool Floral Design Program, is the culmination of the best technical skills presented by the long illustrious list of FlowerSchool Master Florists. This book has condensed the skills and philosophies associated with being a professional florist in New York City into an easy-to-replicate set of instructions that will help you become an accomplished florist yourself. Like a chef planning a great meal or simply cooking a Sunday dinner, you will learn how to select your materials, how to prep your materials, and how to use myriad techniques to accomplish your design goals. The production steps in the following pages are designed to lead you to success every time, allowing you to break free of any nervousness surrounding the process of flower arranging, and to focus on developing your own personal floral design sensibilities.

Florists in New York City often have Herculean tasks ahead of them: from setting up a party for 800 people in under an hour to installing five 20-foot-high

Actor Rob Morgan works to complete an autumnal arrangement at FlowerSchool LA. "All proper ladies and gentlemen need a good florist for sending gifts. Understanding flowers is an important discipline."

arrangements before visitors arrive at the Metropolitan Museum of Art to managing the delivery of over 150 daily flower arrangements through gridlocked city traffic that no mere mortal can navigate. In order to be successful, most of these companies choose to focus on a specialty.

Olivier Giugni is a master of weekly floral installations. It is his main focus, and all of his clients have the same set of expectations. It only makes sense that all of his design work is informed by the tasks that he performs most often in order to make his business successful.

Olivier Giugni Shows Us How It's Done

Olivier Giugni is a master of creating weekly flower installations. It is his main focus, and all of his clients have the same set of expectations: spectacular arrangements delivered on time each week. A master of visual creativity, Olivier's work also requires him to be a master of scheduling and order as well as a creator of gorgeous arrangements.

As the driving creative force behind L'Olivier, the world-renowned floral atelier in New York City, Olivier Giugni has been making arrangements with movement and depth for an A-list roster of private clients for more than 30 years. In addition to adding warmth and excitement to private homes, he has also created stunning florals for the restaurants of Chef Daniel Boulud and for the Carlyle Hotel. Here Olivier tells us a little bit more about his process, and in the following pages he offers a lesson in creating a gorgeous arrangement that accentuates its surroundings, and makes use of whatever materials are on hand.

What makes a flower arrangement modern?

Precise color, shape, scale, proportion, and strong, bold gestures.

What is your style?

Modern flower bouquets need to be simple, architectural, and significant. When I was running Pierre Cardin's flower shop in Paris, the window had to make a statement every week. Often, the window was filled with one-of-a-kind antique vases from Mr. Cardin's personal collection. Sometimes, during couture season, his elaborate handmade dresses would be featured in the window. It was very important that the flowers worked with the window and did not fight against whatever was on display. Therefore, my expression of modern design is not to be loud, but to work in concert with the room or location.

What makes your style unique?

While working in a flower shop environment, you have the opportunity to work with a great many products. Our shop sold lots of tropical plants with very large foliage. So, rather than purchase foliage and fillers from the flower market, I would cut leaves from my palm trees and use them. It makes quite a statement.

Just like any at-home arranger, Olivier sets up his workstation at FlowerSchool with all of the materials he'll need in order to make the arrangement. In this case, we have white amaryllis that have been conditioned (see page 143) and given several days to open, as well as accentuating green leaves and grass that are set off to the right. Now they're ready to be transferred to the rectangular vase in the center whenever Olivier is ready for them.

As he prepares to create his arrangement, Olivier takes time to pluck any petals that are looking wilted or less than perfect. Now is the time to give your blooms their beauty treatment.

As Olivier has noted, scale and proportion are extremely important parts of creating a modern design. In order to make sure the amaryllis are well distributed and that they stay in place, Olivier has created a simple grid using waterproof tape. This tape will not be visible to viewers once the arrangement is complete, but it's essential for creating an arrangement that is orderly and that stays in place.

Now it's time to start adding flowers and accompanying greenery. Here Olivier shows us his famous leaf wrapping technique, where he takes a large leaf and wraps it around the top of the vase. This is just one example of how a personal touch can take an already beautiful arrangement and make it sing.

After adding some additional amaryllis, along with some taller blades of grass for additional greenery, Olivier steps back to consider his arrangement and to determine if any alterations should be made. Notice the ends of his amaryllis have been taped so that they don't split. This is a good reminder to all students that you should constantly be checking your work. Step back often and consider your arrangement. The most successful florists learn to make adjustments as they work.

This is the time to consider all of the components of the arrangement. Is it balanced and proportional? Are there elements of varying height, color, and texture? Do the colors work well together?

After careful consideration and planning, here is the final arrangement. Notice how the underlying grid of waterproof tape is helping to hold all of the flowers, grass, and leaves in place, while remaining invisible to the eye. This arrangement is deceptively simple but requires a master's touch. This technique can be used for all large flowers with long stems such as delphinium or lilies (see illustrations on pages 156 and 154). With practice, you too can learn how to harness proportion, color, shape, and scale to make next-level arrangements.

Advice and Guidance for New Florists from New York City Florist Victoria Ahn

Victoria Ahn creates exquisite parties and weddings for some of New York City's most prestigious clients and venues. Professionally trained in ballet and modern dance, Victoria is now the owner of Designs by Ahn in New York City. She approaches each arrangement like a choreographer, paying close attention to how each color and texture is placed, and how they work together. There is truly no better person to look to for matrimonial design trends, and tips on how to make exquisite table decorations.

What makes a great florist?

You have to have a discerning sense of style. This is something you can't really teach; it can only be found inside. Also, flowers themselves are beautiful. To make them extra beautiful, you must use many different textures.

In addition to her work as the owner of Designs by Ahn, FlowerSchool Master Florist Victoria Ahn also teaches classes at FlowerSchool New York. On this page and the next we see just one example of Ahn's stellar use of color in her work.

ERSCHOOL
Art & Flowers Series

How do you start a flower arrangement?

I start by being open-minded. I don't usually plan. In fact, for a lot of events and photo shoots, I don't even go in with a plan. I am always trying to come up with a new idea, and I am always trying to find the most unique way to create a new look. I go to the market every day to find inspiration, and to search for new colors or new flowers.

Where does your color inspiration come from?

I have always liked playing with color. Museums are obvious places to go for inspiration. I also look at what people are wearing. For instance, when I meet with brides, they don't necessarily know what colors they like, but you look at what they wear, and that tells you what colors they like.

What advice would you give new florists?

Keep it simple. If it is your first time, use just one color.

It only makes sense that this book contains the superb classic styles of FlowerSchool Master Florist Remco Van Vliet, alongside beautifully designed bouquets fit for a queen inspired by the contemporary installation design of FlowerSchool Master Florist Ingrid Carozzi of Tin Can Studios.

Our goal is to have you develop your basic skills with the help of our Master Florists so that you can always make beautiful arrangements, or "living décor," for every occasion. We will show you how to construct a basic set of designs so they can be done easily by you. These designs can be transported easily, the water can be changed, and they are balanced well so they won't fall over once the flowers have drunk up the water. After we get a solid footing on how to use the equipment and skills for making flower arrangements, we will incorporate color using some helpful little tricks found later in the book.

The structures and workflows featured in this book are those used by our own Master Florists. Most are not necessarily appropriate for Instagram or social media, but from flowers in their natural environment to flowers in the dining rooms of the Carlyle, this book features real floral design deconstructed for all. The instructions in this book will provide a foundation for your process and correct any errors you may be making.

You will learn how to properly care for flowers while creating each flower arrangement, making sure not to destroy or overwork them in the process, so they can live their best life.

You will learn how to properly build a vase arrangement. Often, flowers are relegated to a vase that is the wrong size or doesn't fit. That can lead the florist to question whether this flower arrangement "will work." It does take some time to become comfortable with the process and to learn to practice patience throughout the designing process.

You will learn how to extend the life of your flowers as long as possible and within reason. Flowers done right should not die prematurely, but rather they should have a dignified end. Nothing lasts forever.

Most important, you will learn how to manage the flower arranging process quickly and efficiently, so that you can finish the practical steps and focus on the artistic and aesthetic pleasures of working with these gorgeous blooms.

Here, these methods are being deployed by Master Florist Ariella Chezar.

1.

Getting Started

EXIT

DOING FLOWERS IS NOT AS COMPLEX AS IT MAY SEEM TO A BEGINNER. It is technical work akin to cooking. There are some basic procedures you will need to become familiar with in order to make a flower arrangement that you are happy with. Once you learn how to perform these standard tasks, every other decision you make will simply augment the technical work you've done. These artistic decisions are where you'll be able to let your personal taste and vision shine through. For the most part, the artistic decisions that need to be made happen in the beginning stages of a flower arrangement, and at the very end. What happens in between these artistic choices is simply the technical work.

A good comparison would be going to a farmers market at the height of harvest season and purchasing food for a feast. Spread out before you is a huge number of products to bring home with you. What do you choose? Well, most experienced cooks start with a list of ingredients that they will need, while still allowing for the opportunity to buy any item that might look particularly delicious and is in season. The same principle holds true for floral artists: Most will go to buy flowers with a general plan in mind, and specific flowers to find, but one must always remain open to whatever beautiful blooms speak to you when you're shopping. Once you've made your selection at the market, the technical work begins. While a cook will transport everything home, eventually washing the produce and dicing the onions to the correct size, the florist must condition the flowers and prepare their containers. The artistry for both disciplines comes at the very beginning—with picking the materials—and at the very end, when you add a finishing touch of fresh thyme or the perfect basil.

In both cases, the essential first step is to make a shopping list. Having a list will set you up for success by taking the guesswork out of what you plan to make later. When making your list, be wary of anything preventing you from doing flowers well. The first thing that can hold many back, and my personal favorite, is the idea of the "perfect flower." A skilled florist should be able to make do with any flower, and should practice doing incredible things with what you have rather than what you don't.

This is Dutch Flower Line, a wholesale flower business in New York where most anyone can purchase some of the world's most exciting varieties.

As we discussed in the previous section, in addition to buying and preparing your flowers, when you're first getting started you will also need to know how to use your tools properly and how to condition your flowers for proper hydration. You should also have a general concept of your style and a plan for where to place your finished product. More practically, make sure you have a place to work and a garbage can handy. In short, you need to have a good idea of what needs to be done before you can execute.

Every great florist has a streamlined workflow that makes their process unique. For example, some begin with an inspiration, others with using what is available:

> "A great florist is simply a hardworking person who is ready to roll up their sleeves."
>
> —INGRID CAROZZI

> "In my opinion, with color, you have to be daring in order to stand out from a crowd."
>
> —KIANA UNDERWOOD

> "Yes, the acquisition is a thrill, the making of the perfect arrangement is a creative pleasure, but the delight for me is often in giving it away."
>
> —BRUCE LITTLEFIELD

It's extremely important to have your flowers prepped and ready before you begin your arrangement.

A Basic Workflow for Doing Flowers Well

Step 1: Clarify Your Purpose and Determine Your Style

First, you must have a reason for doing the flowers: a dinner party, perhaps? Or maybe a gift for a friend who has been ill? Maybe a perfect harvest from your home garden, or from someone else's home garden, has yielded flowers that are begging to be displayed to their full potential. Taking the time to state your purpose will help you maintain clarity as you work to create your arrangement. For instance, does the arrangement need to last for a week, or is it meant for a four-hour event? Once you've defined your purpose, you'll need to determine the color and style of arrangement that are appropriate for your vision. See Chapter 3 for more information on choosing flower colors and arrangement styles.

Step 2: Gather Your Materials

Once you have defined your reason for getting to work and selected your style and color scheme, your next step will be to gather your materials, whether from a flower market, a grocery store, or elsewhere. In addition to your flowers, this is the step when you should confirm that you have all the tools you need, and that you have the correct vase, as vessel options vary widely. This is the moment when many florists become overwhelmed. After all, there are over a hundred different species of rose and thousands of hybrid varieties. And that's just one flower! However, your options as an at-home florist will be more limited, and therefore, more manageable. We will discuss acquiring materials in detail later in the book (see Chapter 4).

The first step in the flower arranging process is to stop and consider your reason for creating an arrangement. Do you need a centerpiece for a dinner party? Or perhaps a bouquet for a friend? Define your purpose, and let that be your guide.

Step 3: Hydrate Your Flowers

Once you have gathered your materials, you will then need to condition, or "hydrate," your flowers to get them ready for use. Hydration times will vary depending on the flowers you choose. We will discuss this step in more detail on page 143.

Step 4: Adjust Your Vase If Necessary

This is your last chance to make sure you have chosen the correct vase. Take this opportunity to look over your vase and your materials to confirm that they are an ideal match. We'll discuss vase selection in more detail in Chapter 4. Please refer to this section before you get started.

Step 5: Get to Work

Finally, it is time to use your floristry skills and make the flower arrangement. It is no easy feat to get all these steps to line up perfectly every time. There are also tons of distractions along the way. A glaring example is the incredible beauty that you will be confronted with on a frequent basis. These peonies are a good case in point! But through years of experience I've acquired some tricks of the trade that will help guide you and ensure that you make the perfect flower arrangement.

Here we see a parcel of unopened blooms. Choosing the correct stage to buy your flowers will depend on how you plan to use them. Are they going to be a gift, or will you be using them for an event? If you're buying flowers as a gift, you should purchase tightly closed flowers that will open over time. If you're designing for an event, the flowers must be open already. Once you've purchased your flowers and brought them home, it's very important to hydrate them and get them ready for use. See Chapter 5 for more details.

Using the Equipment

For all flower cutting, it is extremely important to cut slowly, cutting no more than a quarter inch at a time. Before cutting, stick your flower into the vase you've selected to see how much it should be cut. Once you've made your cut, insert the flower again to confirm you've reached the appropriate length. Repeat this step over and over again until the flower is the perfect height. Go slowly, and be patient. You can always cut off more of a flower's stem, but once you've cut it too short, you can't add the stem back. Be sure before you cut!

Floral Knives

Not all flower stems need to be cut with a knife, but rather than try to determine which blooms will benefit, just use a knife to cut them all. The flowers that will benefit will bloom more happily. The flowers that won't benefit won't be any worse for wear. To begin, cut one stem at a time. Once you become comfortable, cut 10 stems, then 20. Watch how quickly you can get your flowers cut and into water.

Long Scissors

These are perfect for sneaking into an arrangement as it's coming together and taking out a poorly behaving leaf or flower at its base without disturbing the larger design.

All-Purpose Clippers

Generally, all clippers must serve the same function for florists: to cut stems! However, after years of using these tools, most florists find that each performs a specific and unique function. These small pruners are a staff favorite at FlowerSchool because they are sharp and easy to hold.

Wire Cutters

As the name suggests, wire cutters are essential for cutting wires. They're also good to have on hand so you don't have to use your expensive pruners for cutting tough wires and chicken wire.

Cleanliness

Your tools should be cleaned before and after each use in order to ensure that you're not spreading rotting plant matter and bacteria around your workspace, and so that you're not a slob. Simply use soap and hot water for all.

Staplers

You'd be surprised how often a stapler comes in handy, especially when coning your flowers to help them stay straight (see page 151).

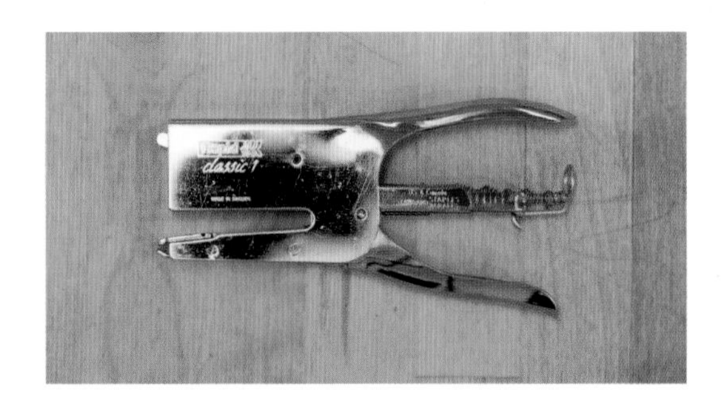

Large Loppers

These are the largest type of manual garden cutters, and they're meant for use in cutting sturdier materials like twigs and branches. If your arrangements include flowers foraged from your garden, these may come in handy.

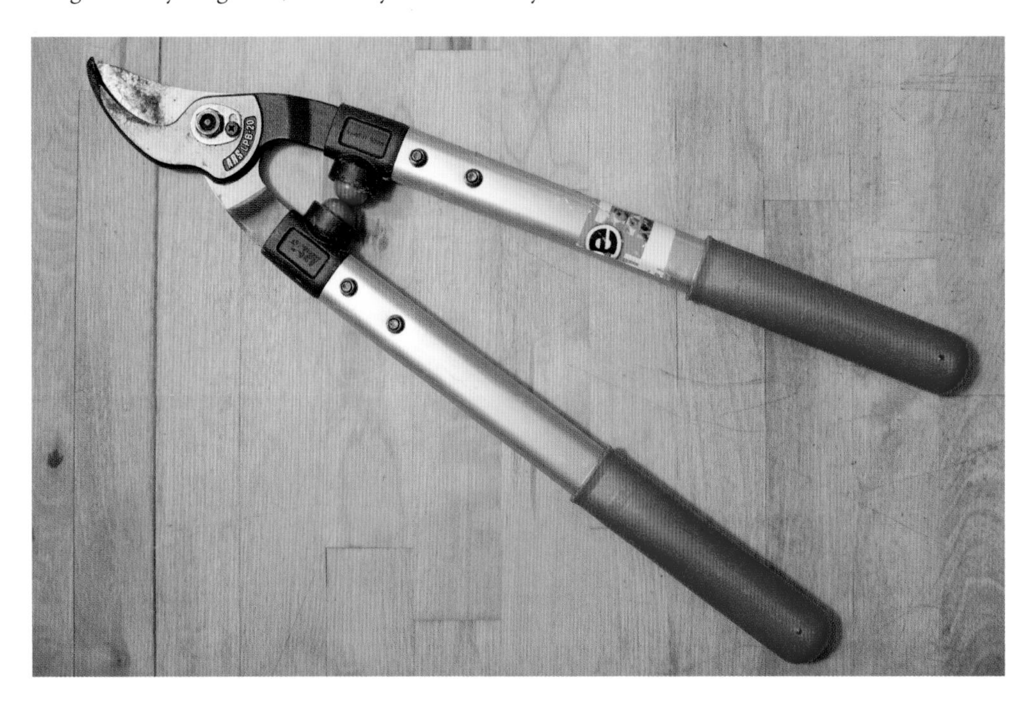

> These tools are sharp, so you better be sure not to cut your finger off.
>
> —OLIVIER GIUGNI

Other Items to Have on Hand

Chicken Wire

This thin, flexible steel wire is perfect for keeping your stems in place. Make sure the brand of wire you use is coated so that it doesn't rust and contaminate your water.

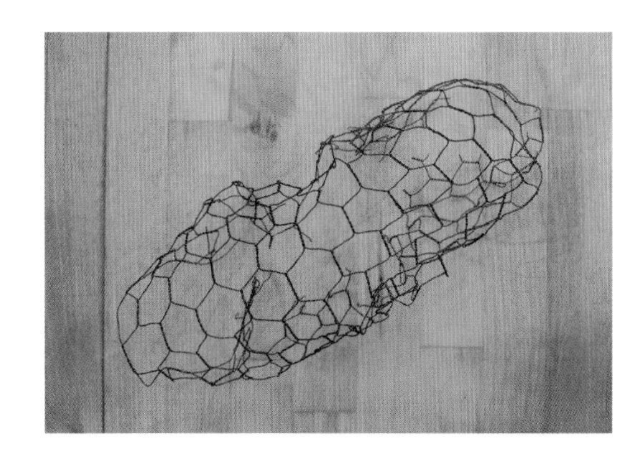

Waterproof Tape

This is the best product to use for tape gridding and holding stuff down. It's crucial that items be firmly taped before water is added to any vessel, or the tape just won't stick. The staff at FlowerSchool prefers to use Oasis brand waterproof tape for most projects.

Rope or Twine

You'll often find yourself needing some rope or twine to tie flowers together. Best to have some on hand at all times.

Angled Cuts

There are several reasons that flowers should be cut on an angle. First, an angled cut will increase the exposure of capillaries in the flower's stem to water, which allows the flowers to drink in the water more efficiently. Second, when putting the flowers into a vase or bucket, cutting on an angle will ensure that the flower ends are not blocked by the bottom of the bucket. Finally, when making actual arrangements, you will find it much easier to insert stems cut at an angle into a crowded vase of flowers.

The angled cut is an important skill to have in your repertoire as you start to work with flowers. Here (above and right) we are conditioning roses by making angled cuts with a pair of clippers. Later, we will demonstrate how to complete this process using a knife.

2.

Choosing Your Colors

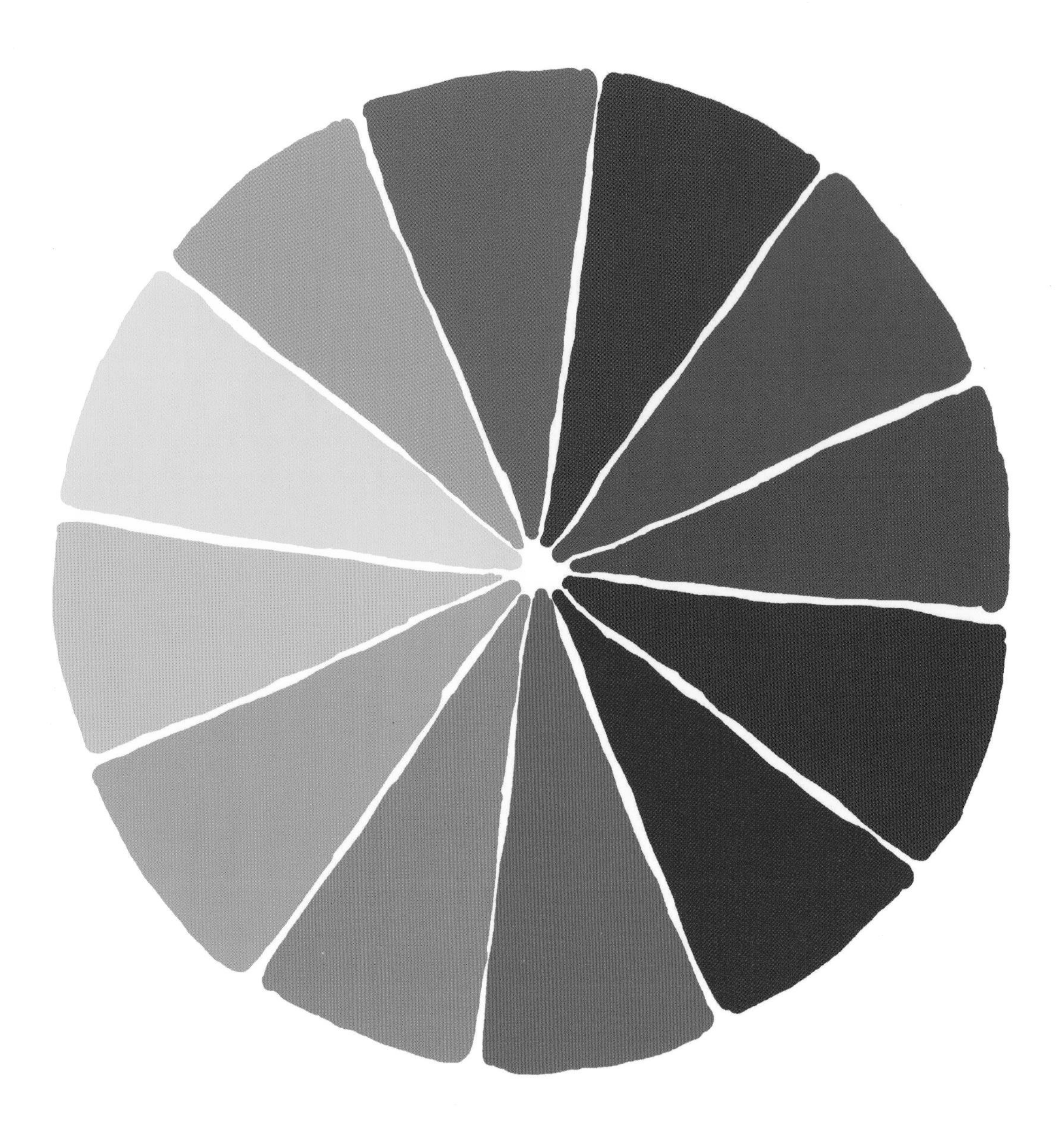

A color wheel is an essential tool as you begin to consider what colors you'd like to use to build your arrangement. There are 12 colors on the wheel, which are broken into three groups: primary colors (red, blue, and yellow), secondary colors (green, violet, and orange), and tertiary colors (blue-green, blue-violet, red-violet, red-orange, yellow-orange, and yellow-green).

FLOWERS HAVE THE UNIQUE ABILITY TO TAKE ON AN UNLIMITED AMOUNT OF COLOR. Their variations in color are often extremely subtle and may not be visible to the naked eye, but a flower's color is indistinguishably linked to its life force. The color of a flower is nature and life itself, held in the palm of your hand. If you are anything like me, when you walk into a flower market, wholesaler, grocery store, or even a neighbor's yard, you immediately lose your train of thought and get distracted by all the color and nature that surrounds you. The eye naturally gravitates toward the most incredible flowers available, so much so that shoppers confronted with beautiful blooms may completely forget what flowers they were originally looking for . . . as well as their budget and any friends they may have brought to the store with them.

While color is one of the things we love most about flowers, it can also be the most divisive. Some people simply hate yellow. They never want to see it. But, at the same time, these haters of yellow might love a light cream or mustard. It can be difficult to come up with a pleasing color palette that appeals to everyone. Most florists tend to start small, choosing certain hues or analogous colors. My own recommendation is to start with just one color.

When determining color style or theme for an event, you must be careful and prepared to choose wisely. Often, color choice can be the most important decision you make as a designer. It can trump all other design elements. One misplaced color can set the eye ablaze and ruin hours of thoughtful work and design. Don't fall victim to the wrong color choice!

Important Words to Know

Before getting started, here is a quick glossary of terms that will help you determine how to choose specific colors. Ultimately, as a florist or at-home floral arranger, you will need to use your intuition and personal preferences to decide what color flowers you'd like to work with, but the following list of terms and concepts will help guide you as you determine what color or colors are right for your arrangement.

Color Range

Flowers that exist within the same color range are referred to as "analogous" and are exactly, or almost exactly, the same. Designs that feature flowers from the same color range are designs made up of one basic color. This is helpful for more modern designs when blocks of color are needed.

Color Hue

A hue is a specific gradation, or shade, of a color. There are 12 hues on a typical color wheel, from which all other colors on the spectrum may be derived. Of the 12 colors on a color wheel, there are three primary colors (red, blue, and yellow) that can, in theory, be mixed to create all other colors. By mixing two primary hues together, you can create a secondary color. The three secondary colors are green (a mix of blue and yellow), violet/purple (a mix of blue and red), and orange (a mix of red and yellow). By mixing one primary color and one secondary color, you can create a very chic mix of colors. These are known as tertiary colors, and they include blue-green, blue-violet, red-violet, red-orange, yellow-orange, and yellow-green.

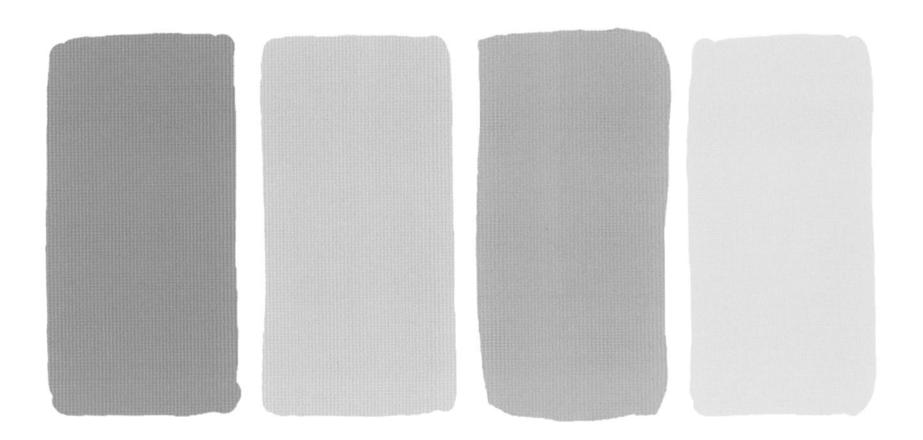

Shades and Tints

Shades and tints are color variations created by mixing one of the 12 colors on the color wheel with either black or white. The reason shades and tints are dissimilar from primary colors or color hues is that they are variations created using only one base color, which is then mixed with either black or white.

Contrasting Colors

Contrasting colors appear on opposite sides of the color wheel. Arrangements made with contrasting colors tend to be very busy and can often backfire if the appropriate mix isn't chosen. We see a lot of contrasting colors used on celebratory balloons or to represent sports teams. It's best to match your contrasting colors with other colors that are present in whatever environment your final arrangement will be living in, whether at home or at an event.

A How-to Guide for Picking the Perfect Color

In our experience, most beginners simply go to the market and pick all their favorite items in the colors that they like. Unless you have years of experience under your belt, this is a rookie mistake. Most first-time arrangers make the mistake of thinking that because they are buying flowers in colors that they like, all the blooms will somehow magically go together. Nothing could be further from the truth! All well-executed design motifs need a color palette. To determine what colors to use for a party and, subsequently for your flowers, we offer the following basic starter guide:

Step 1: Start by Recognizing Some Seasonal Palettes

Pick one or two colors that match the colors of the room where the flowers will live. You don't have to be too specific here. You can choose colors that complement the season viewed through the room's windows or that pair well with a painting on the wall. Traditionally, each season has its own set of corresponding colors, and this is a nice place from which to start understanding how to design seasonally. Below is a list of some basic starter palettes based on season:

- Summer—yellows, greens, browns, and light blues. Summer flowers are typically field flowers that can be picked along the side of the road or in a field on a late summer day. Best arranged in a mason jar, or a similarly informal vessel, summer flowers carry an air of nonchalance. A summer arrangement should look effortlessly thrown together. After all, it's summer!

- Spring—pastels. Spring is a time for rejuvenation. Spring flowers don't often carry rich colors, but veer more toward softer tones and pastels. You won't find brown in spring, but you will find lavender; and no red, but pink instead. These colors all tend to go well together and should be combined with the joys that spring always brings. The above image is an example of a palette that's appropriate for later in the spring, as opposed to an early-spring palette, which is much more pastel heavy.

- Autumn—rich browns and reds mixed with creams. Autumn is the traditional harvest season of the Northeast United States and the change in weather from the warmth of summer to the cool air of fall always brings out the leaf peepers. Autumn's rich colors mark the turning point from summer's green foliage to the rich tones that represent the beginning of nature's seasonal respite.

- Winter—frosted reds, deep reds, and crisp whites. The cold weather always brings with it late, dark nights of celebration and relaxation. Dark reds if you have them, or winter whites mixed with metallics are the perfect way to set a winter mood.

Step 2: Choose a Style Using Color

When you are choosing a color palette for a party, there are certain palettes that are viewed as opulent and romantic:

ELEGANT

Elegant color palettes tend to be pastels in a range of different hues, often a mixture of light colors such as cream, lavender, pink, and shades of white. The colors for elegant palettes would all be at the lighter end of the paint strip, and they invoke an ephemeral, romantic, and dreamlike quality. Stay away from loud colors and jewel tones such as burgundy and purple.

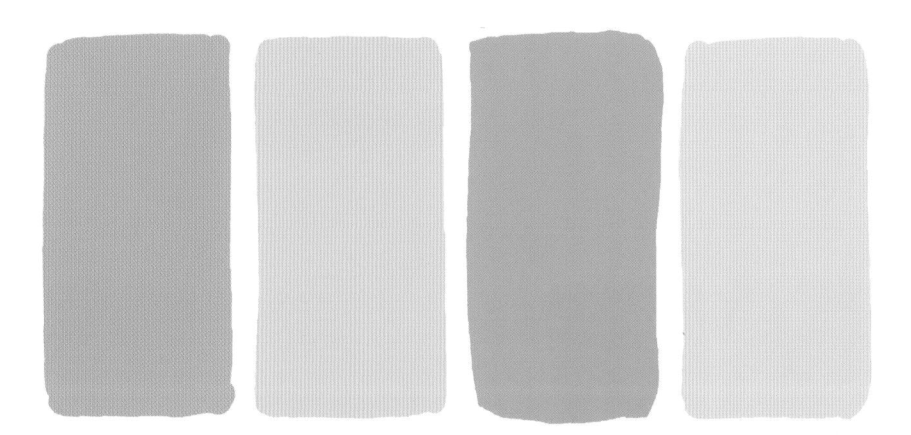

RICH JEWEL TONES

Rich color palettes are meant for autumnal parties or when you need to display bold pops of color. These seductive colors can overwhelm viewers, especially when expressed in flowers. The colors for rich palettes would all be at the darker end of the paint strip, and they tend to look really great when paired with gold and other metallics.

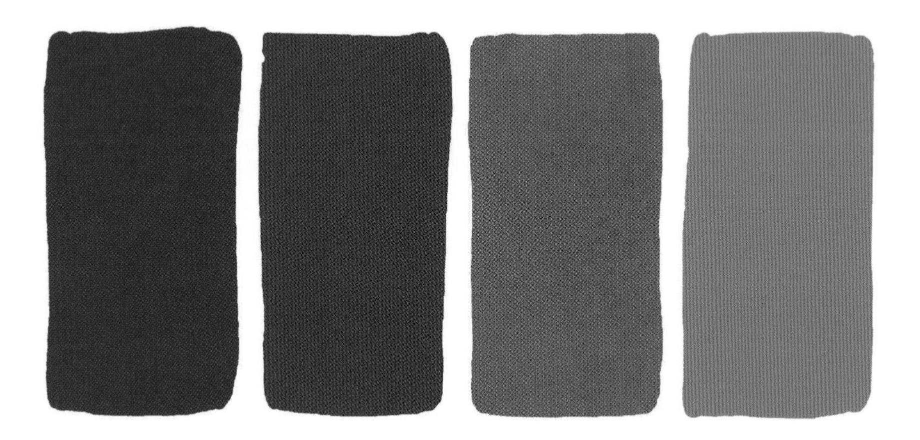

WOODLAND

Woodland palettes include a mix of light and dark earth tones. Throwing in a bit of birch bark, moss, or other organic material doesn't hurt. The woodland vibe should remind the viewer of a walk in the forest. The colors, textures, and even the smells of your arrangement should replicate a long walk in the mossy woods.

GARDEN DESIGN

What you think a garden should look like will determine what makes the garden design palette uniquely yours. One common thread, however, will be lots of shades of green. Be careful not to use too many different shades or your garden design palette won't look natural.

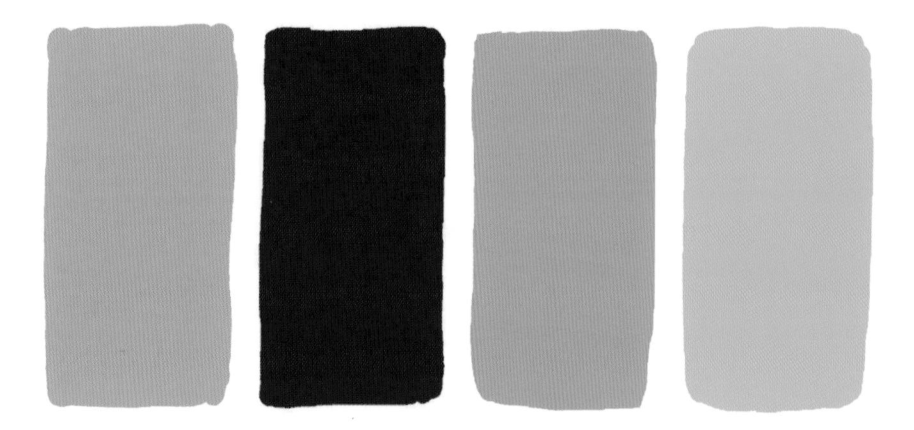

Step 3: Choose Your Supporting Colors

Once you've picked your main color palette or "color story," it is helpful to find supporting colors for your story. Using a color wheel, find the strip that most closely matches the tone of colors you've chosen.

- For a more modern aesthetic, find flowers and materials that are all on the same color strip.

- With red as a secondary color, the entire arrangement can now become less contrasting and more chic. Simply fill in your design with colors that appear between yellow and red on the color wheel.

- For a more garden-centric approach, use six different colors and a variety of foliage. Foliage is usually the easiest thing to find, so be generous when working in the garden style.

- For a more elegant style, use two of the central colors chosen from below. As a general rule, elegantly styled flowers tend not to include green foliage.

The Basic FlowerSchool Philosophy for Building a Style and Color Story

Flowers tend to be awash in an array of colors that run the gamut of primary colors and their derivative hues. When picking colors for your flowers and vases, there are several things to consider. First, the color of the table, or tablecloth, on which your arrangement will be placed. Also important to consider: the color of any plated food that might appear on the table and the color of the room in which the arrangement will be placed. As much as possible, you should try to build an arrangement that complements its surrounding environment.

For the sake of this exercise, let's build an arrangement using only one color.

When making an arrangement in a traditional style (more on different styles of arrangements in Chapter 6), we can start by simply choosing one color of flower and flower shape, and pairing them with a vase of our choosing. Then, as we add texture to the flowers and the surrounding accoutrement, we can see clearly how much impact a seemingly simple change can have on the overall look. In the first image above, we've chosen a green flower with a white-and-gray vase and we've paired the arrangement with a white cloth. All of the flowers are shoulder to shoulder.

To make the design a bit more contemporary, we'll next add an analogous color: cream. By filling in the arrangement with a variety of shades of cream and green, we've given the arrangement additional levels and we've given the viewer more to look at.

In the final image, we've added additional, non-floral greenery of various heights to fill in the entire arrangement and make it pop. These variations could continue with several different color palettes. The options are endless!

PRO TIP: When using two colors, you must be careful to avoid making a hot mess! If you're working with colors that are on opposite sides of the spectrum, try not to include too many shades. Remember: there is no such thing as bad flowers, just bad flower choices.

Kiana Underwood, the Owner of Tulipina, Discusses Her Approach to Color

Kiana Underwood is the owner of Tulipina, a luxury floral design and event studio that is known around the world for its playful and inventively colorful arrangements. Her unique approach to color, coupled with her inspired floral selections, draws admirers and floral designers of all stripes to her sold-out FlowerSchool workshops. In addition to amassing an enormous Instagram following and authoring her own successful book, *Color Me Floral*, her work has also been featured in publications as varied as the *New York Times*, *Elle Decor*, *Country Living*, and *Town & Country*. Luckily for us, Kiana has kindly found the time to answer a few questions about her ingenious knack for mixing style and color, and she has provided us with a few images of her gorgeous work to serve as inspiration.

How do you get started in designing and making a single flower arrangement?

I generally think about color first and how to position my ingredients in my vessel so that each bloom is shining. Shape and texture also play a great role in what makes my arrangements.

How do you find your inspiration?

My inspiration often comes in the moment, without much preplanning. As I see a design take shape, I know how I want to continue and, ultimately, complete it.

What about finding inspiration when it comes to color? Does that happen in the moment, or do you plan ahead?

In my opinion, with color, you have to be daring in order to stand out from a crowd. For me, color inspiration comes from the flowers themselves. If you look at a flower closely, you will see that it is made up of many different shades and colors. For example, a flower that appears orange as a whole may have streaks of magenta and dark purple inside, and to combine those colors in an arrangement can be quite magnificent. I know I have achieved my goal when my arrangements stop people in their tracks and I hear them say, "Wow, that color combination! I would have never thought about those colors together…but it looks amazing!"

3.

Buying Flowers

FLOWERS COME IN A MULTITUDE OF SHAPES, SIZES, COLORS, AND HEIGHTS, AND EACH VARIETY CAN BEHAVE IN DRASTICALLY DIFFERENT WAYS. Some flowers lean, like tulips. Some are just as straight as can be. Some flowers are flat, like hydrangeas and roses. Finding the appropriate combination can be a challenging prospect. It's enough to make even the most experienced florist's head spin! If you're feeling overwhelmed, know that you are not alone. You may go to a flower market, or to a professional florist's shop, and see someone with experience pulling flowers seemingly without thinking. "I'll take those buckets, and those flowers. *All of them*," they say, but how do they know? The answer is a combination of wisdom and planning. There is a certain amount of floral wisdom and intuition that can only be gained through experience. Over time, you learn what flowers you like best and what flowers work for specific occasions. But planning is something you can learn right now that will help you immensely. I know I've said it before, but it bears repeating. Always have a plan! Try as hard as you can to know what style of arrangement you would like to make, and what flowers and materials you will need in order to make your arrangement look the way you like. This may seem like a daunting task at first, so we've provided a guide to help you choose.

Step 1: Choose a Style

How many stems you will need to create your desired arrangement is the fundamental question of floral design, and it's also one of the most difficult to answer. For florists, the answer is usually "How many can I get for the money I've budgeted?" However, when trying to plan, basing decisions entirely on budget isn't always helpful. Life is what you make of it, and flower arranging is no different. You have to play the hand you're dealt. Even if you have all the money in the world, you could still end up at a flower market that doesn't have anything that strikes your fancy. And if you only have 10 dollars to work with, you should still be able to make an arrangement that looks beautiful. It all comes down to how you use the materials you have. The first step to take is to choose a style. What follows is a general list of major floral style groups. As you grow more comfortable with your design skills, you should feel free to build on these concepts and make them your own.

MODERN

This style tends to use only one color or one flower. When people refer to modern design styles, they typically mean something minimal, architectural, or graphic. Think traditional, straight forms that include little to no foliage. This low, modern rose cube arrangement is perfect for a dinner table or a coffee table. It is built in a glass cube that is 4 inches tall, 4 inches long, and 4 inches wide, and it holds 25 roses.

GARDEN DESIGN

Garden-style flower arrangements tend to include many different flower varieties and foliages. A typical arrangement in this style will include at least six different varieties of flower plus four or more types of foliage. For this style, it's important to remember that the foliage is just as important as the flowers, and each element must be given room to be viewed.

ELEVATED AND GRAPHIC

Meant to be viewed from the side, this style is perfect for a bar or entryway table. Flowers are typically not closely tucked together and the stems used are tall. Flowers that look nice in tall vases like this are Eremurus, flowering branches, and delphinium. And if you have the correct vase—one with a small opening—you typically need only 5–10 stems.

CLASSIC

Most often, professional florists make so-called classic arrangements by holding the flowers in their hands and hand-tying them. Think of a standard bouquet. You should have 30–35 stems for starters and they should include a mix of flowers in varying colors, all placed in a simple vase. A classic arrangement is the type of arrangement for which the old adage, "Flowers should be one-and-a-half times the height of their vase" was coined.

CONTEMPORARY MODERN

Typically, elegant arrangements are made in shades of whites and pastels with little to no foliage, and they're often arranged in standard symmetrical forms. Think of a round ball of hydrangeas and not a stem or leaf in sight. Flowers should be fairly straight and shoulder to shoulder, which will help them look as expensive as possible. This arrangement is one step removed from the traditional modern design in that it includes more than one color and the stems are hidden. It has a strong uniform shape and multiple textures. This design can easily be achieved using a variety of flowers that are similar in size and shape, including hydrangeas.

GARDEN

This is a good example of a garden style arrangement, which typically includes three or more types of green foliage that are mixed with an analogous color palette. The loose and varied foliage makes it a garden design and all the blue/purple tones make it analogous.

Step 2: Choose a Flower and a Size

Once you've chosen a style, it's time to decide which flower varieties will give you the look you're going for. While we make every attempt to guide your flower procurement and help simplify your design decisions, there are exceptions to every rule. For example, while a mix of three roses, three hydrangeas, and three delphinium will typically look peculiar, it is not impossible on occasion to create a trio that looks spectacular. If you up the numbers, say, 25 roses, 25 hydrangeas, and 25 delphinium, and use the correct design mechanics, you will have a much easier time being successful. However, in this book we are interested in helping the casual, at-home florist create standard vase arrangements, so we've necessarily narrowed the creative margins in order to ensure success for everyone.

HEAD SIZE

This is most evident and useful with hydrangeas. Head size is measured by saying 10+, meaning that the head is 10 cm or larger; 15+ is 15 cm or larger; and so on for 20+, 25+, 30+, and larger. Placing a huge, 30+ hydrangea next to a rose will make the rose look puny. It's better to mix the larger hydrangea with a cluster of roses.

STEM LENGTH

Typically, the longer the stem, the more robust the flower head and the better the quality. They start at around 40 cm, which is typically grocery store–grade, up to 130 cm, which are typically only special order. In our experience, 50 cm, 60 cm, and 70 cm stems are typically sufficient for professionals. For some varieties, a shorter stem is sufficient.

FLOWER FACE PROFILE

Some flowers have a relatively flat profile or face, such as roses and hydrangeas (both mentioned above). These two flowers can easily mix because they are similarly shaped. There are varying degrees of flower shape, which is helpful to keep in mind. For example, gladiolas and delphinium are very tall, making them difficult to mix with roses and hydrangeas.

Step 3: Pick Your Vase

We always want your arrangements to give a sense of abundance and to be spilling out of the vase. So, as a general rule of thumb, you should pick the smallest vase you can get away with in relation to the size of bouquet you would like. You want your flowers to appear as though they're overflowing out of your vase, rather than being swallowed by it.

If you want to give your arrangement some height, you will need your vase to be higher than it is wide. For instance, a vase with an opening that is 4.5–5.5 inches wide should be 7–8 inches tall or taller. A vase of some height is especially good for tall flowers like delphinium. For a more in-depth discussion of vases, see Chapter 4.

> **PRO TIP:** If you don't have access to good vases, only simple boring ones from a secondhand store, you may want to find some materials to help "craft up" your vase. Birch bark or fabric is usually a good start.

Step 4: Where to Buy

Materials and their uses are a central part of any florist's day-to-day working life. Below is a list of the most common places that florists go to get materials. You may be surprised by some of the locations on this list because most plants are hiding in plain sight!

Depending on where you're located, you may or may not have access to "premium retailers" and, therefore, you may have to dig deeper creatively in order to work your materials to their maximum effect and express greatness. Fear not, this is a blessing in disguise—and a chance to get creative! Look in places that most wouldn't ordinarily consider. Take note of the natural world around you, and see what flowers and foliage are ripe for the plucking.

GROCERY STORES

Grocery stores are excellent places to get product, if you know how to order correctly. Often, your local grocery store will have a flower section or a few buckets of flowers for you to choose from. Reputable chains like Trader Joe's and Whole Foods Market usually have a nice flower program, but they are not alone. If your local grocery store has a flower program, we recommend you set up a meeting with

Flowers from your local deli or grocery store can be used to produce arrangements that are every bit as gorgeous as those made with designer blooms. All it takes is a basic understanding of how to work with flowers, how to choose the best colors—in this case, green and white—and a little touch of personal creativity.

the buyer or person in charge of ordering. You can find out when their flowers are typically delivered and what type of flowers the store has access to. It's an unfortunate truth that flowers delivered to grocery stores are often left to sit in boxes in the back of the store and basically die on the stems. It's also common for untrained employees to hydrate the stems incorrectly, which can drastically cut short a flower's life. Knowing when flowers are delivered will allow you to pick them up as early as possible, ensuring you get the maximum amount of flower life in exchange for your time, effort, and money. Thus, the sooner you buy your flowers, the better. Don't be afraid to approach your local grocery store. After all, if they play their cards right, you'll become a repeat customer, buying flowers from them before they even hit the shelves. They'll love you, and you may be able to help them expand their business!

If you're pressed for time and you want to purchase flowers that are already on display in the store, I recommend *only* purchasing display flowers that you know are fresh because they don't look like mush. Oftentimes, the paper that these flowers come wrapped in can disguise decaying stems beneath. Make sure you look under the paper before buying!

For table displays, I find the real value in getting materials from your local grocery store actually comes from the produce section. Think fruits and vegetables! A string of grapes or tomatoes on the vine can really enhance a compote vase, Dutch still life, or horn-of-plenty decoration.

FLOWER MARKETS

Flower markets are where most professional florists purchase flowers for their businesses. By virtue of the popularity of florals in art, fashion, culture, and on Instagram, many professional flower markets have also become a bit of a scene—a chic place to see and be seen, with the likes of Putnam & Putnam showing their wares. Despite the hubbub of attention, a flower market is no different than any other commodity market in that it involves the buying, selling, or trading of raw product, similar to less sexy markets that sell cacao beans, fruit and vegetables, or even gold. A professional flower market often includes many different sellers who are each offering literally thousands of varieties of flowers in almost every color imaginable. Don't let this intimidate you! A wholesale flower market is not much different from your local grocery store; it's just operating on a larger scale and offering a wider variety of product. The same techniques you use when buying from your local store apply here. You'll want to find out what kind of flowers each vendor has access to and when they are delivered to the market. And, as always, make a list of what you want to buy before you go! There is often this giant sucking sound that flower lovers hear when they enter any well-appointed flower market. That is the sound of money being sucked out of your bank account! As you confront row after row of incredible flowers, the beauty can become overwhelming and you may find yourself overspending.

The number of flowers available for purchase at professional flower markets can often be overwhelming. Making a list of what flowers and colors you want before you head to the store is always a good idea.

HOLEX

Words of Wisdom From Gary G. Page, Flower Market Stalwart

Don't take it from me—let's hear what Gary Page of G.Page Wholesale Flowers, a fixture of the 28th Street Flower District in New York City, has to say. Over the years, when we've visited the market with classes from FlowerSchool, Gary has often taken a break from taking inventory and checking product quality to give a tour of the market. If you're ever lucky enough to be a part of one of our Flower Market Tours and you run into Gary, you'll undoubtedly hear something like this: "You know, folks, I've been doing this for a long time and I've seen New York's next great designers come and go. But the amazing thing is seeing people who are organized, who arrive with a list of materials, and then shop according to what their client's needs are. Ten years later, they're still at it, making a good living and having a wonderful time. Then there's the other group. The people who show up without knowing exactly what they need, and they end up purchasing things they can't afford. They're frazzled and always a day late, a dollar short, and unfortunately miserable. Don't be that second type of person!"

PRO TIP: For bigger markets, this can be a completely overwhelming experience for a novice florist. During tulip season, for example, there can be over 200 varieties to choose from. Asking for a white tulip can be complicated; there is white single, white double, white parrot, white frilled, as well as a white French grown in France, a white French grown in Holland, and a white French grown in New Jersey.

Scarlet

Even more flowers at G.Page Wholesale Flower Market. If you were to walk into a market such as this without a specific list or plan of action, you might easily find yourself leaving with a mishmash of blooms that don't work together, no matter how beautiful.

The reality of shopping for flowers at a farmers market can often be quite different from the idealized version most of us have in our heads. You can never be sure what they'll have in stock, so be ready to go with the flow and to alter your plans on the fly, if needed.

FARMERS MARKETS

A farmers market is the traditional way every flower lover hopes to experience purchasing loose blooms for the first time. Much like shopping for a perfectly ripe heirloom tomato by scouring the stalls of an open-air market, searching for flowers via face-to-face conversations with local farmers can be a source of great joy for passionate flower fans. A farmers market offers a fun, casual way to pick up exciting blooms and to meet directly with all kinds of growers, ranging from local farms with fairly sophisticated production capabilities to amateur botanists working more for passion than profit.

As fun as they are, farmers markets are always something of a hit-or-miss proposition since many flower farmers who sell in these locations are cutting from the supplies they have ready that same day. While you can be certain that the flowers you're buying are fresh and local, there's no way to be sure that they'll have the quantity you need. Furthermore (and in my own personal experience), some farmers have a tendency to cut their flowers too late. If you're buying flowers that have already opened, like lilies, you can be fairly certain that they won't travel well or last very long. Buyer beware!

PRO TIP: If you find a passionate grower or amateur botanist, it's worth striking up a friendship. Oftentimes, these growers charge higher prices because of the tremendous amount of time and effort they put into their craft, but just as commonly, they'll offer deals to others who share their love of flowers.

A trip to the flower market is always a pleasure, but making a list before you go will make your life much easier, and it will ensure that you leave the market with everything you need, and nothing more.

Asters
Dalhia
8 00
10 stem
$5

FORAGING FROM YOUR GARDEN . . . OR ELSEWHERE

One of my favorite places to get materials is right outside my door. Not a day goes by when I don't see a flower or tree and wonder, "What is the street value of that?!" or "Wow, the possibilities!" or "Look at the shape of that thing. I would love to use it." One of a florist's favorite pastimes is trying to make something out of what appears to be nothing—our very own way of sculpting from a stone.

Once you have settled on your color scheme and your vase, everything becomes a possible material to use. This sense of freedom can really open up the flower arranging process and make it exciting.

When cutting from a garden, there are a few tips for making sure that the flowers will last as long as possible. First, gardens can be filled with bacteria. It will be helpful to use a HydraFlor or FloraLife Quick Dip product (see the hydrating section on page 143 for more information). Second, it is very important that you cut any flower or branches first thing in the morning, preferably after a good watering from either a garden hose or rain. That way, the flower has stored up a lot of food and hydration.

Once you learn how to forage for and cut things, the less need you will have for the traditional flower markets and grocery stores. Your flower-sourcing process can become part adventure, part scavenger hunt, and part nature hike. We always recommend that florists spend more time outdoors, seeing how flowers and foliage exist in nature. Foraging for flowers and pulling items from your own garden is a great way to enhance your eye for detail and design.

PRO TIP: Make sure that any items you forage are cleaned properly, paying special attention to removing any bugs. Yes, bugs! One of the joys of nature! A simple technique for removing bugs without getting your hands dirty, shall we say, is to submerge the flower, moss, ivy, or other foraged item completely under water for a few seconds. The bugs usually just float to the surface.

PRO TIP: When borrowing from a neighbor, be sure you get permission. When it comes to civic pruning, say from a nearby park or a hiking trail in a state park, it is important to know what your local laws say regarding cutting. Familiarizing yourself with these laws will let you know where and when it's safe (and legal) to cut.

PLANT NURSERY

Your local plant nursery, or the nursery department at a nearby Home Depot, can often be your best bet for finding seasonal, locally grown varieties of flowers and plants. From potted bulb flowers in the spring to mums in the fall, there is almost always something fresh that can be purchased, or even cut directly from a plant (see page 197), and put to use. Also, these cut flowers don't need to be refrigerated, just watered.

PRO TIP: Knowing when a bulb is going to bloom is crucial. For example, when you purchase a planted amaryllis bulb, you would know a bloom is forthcoming if the bud is cracking and you are beginning to see color. However, you may still have to wait an entire month to actually see the flowers. Each bulb blooms at its own pace, and you should plan accordingly.

FOLLOWING PAGE: *Plant nurseries, such as Caribbean Cuts Corporation in New York City, are a great resource for fresh, local blooms, no matter the time of year.*

ONLINE

Shopping online offers many prospective buyers the chance to shop for flowers in a low-pressure environment, which can be a real gift. Purchasing flowers online is also becoming increasingly popular in hard-to-reach places. As convenient as online shopping may be, I must stress that when you buy flowers online, you are at the mercy of whatever shipping company is in charge of getting the flowers to your door, and some companies are better than others at protecting the quality of the product. Whenever possible, it is best to use a wholesaler, as they will be able to replace any damaged blooms, and they'll have a better workflow for managing this process quickly. Also, be wary of delicate flowers that don't ship well, like dahlias. Stick to tulips and roses.

Practice at Home

Let's say you're planning a party and you've decided on red as a theme color. Pay a visit to your corner bodega, local grocery store, or nearest Walmart and only purchase flowers that are red. You can choose any hue that you like, as long as it derives from red. The same goes for size and shape. Choose any you like. Whatever you choose, be sure to get a few bunches. Once you're home, condition your flowers (see page 143), and put them together in an arrangement. As long as you've stuck to your color scheme and chosen the correct vase, you should have a great piece of work. Repeat this exercise using two colors that go well together, like purple and white. Practice arranging the two different colors together in a bouquet that you find pleasing. I've suggested purple and white, but it will not make a huge amount of difference which colors you choose. As long as they're complementary, you will end up with a nice design.

4.

Choosing a Vase

IF YOU WANT TO DO FLOWERS WELL AND FIND A SENSE OF SATISFACTION IN KNOWING THAT EACH BLOOM HAS BEEN GIVEN THE CHANCE TO LIVE ITS MOST PERFECT LIFE, THEN YOU WILL NEED TO HAVE A MULTITUDE OF VASES ON HAND. If done right—and for lack of a better term—you must become a "vase collector." Not only do you need a good selection of vases, but you'll need them in many different shapes and sizes (5 × 5, V-shaped, tall, short, etc.), and it would be wise to have at least two of each. Everyone enjoys a good pair of vases. Why?! Because having the perfect vase is like having the perfect pair of shoes to go with your outfit. It makes *all* the difference.

There are hundreds of different types of vases out there and it's a great idea to have a plentiful collection of them if you are going to be able to take advantage of all the beauty that nature and life will give you. Early on in the process, when gathering materials, you may not know what will strike your fancy until you see it. So, when purchasing flowers or coming up with concepts, keeping an open mind is paramount. You should have a list and a general concept in mind when you go shopping, but it's not unheard of for florists to get the overwhelming desire to purchase a flower that's not on their list because of its striking beauty. Maybe you thought roses were what you wanted, but then you settled on some stunning gladiolas—and now you have to reimagine what vase to use. Having a large vase collection will give you the flexibility to purchase whatever flowers you like. When you determine what type of flowers you want to work with, and what the appropriate vase is to house them, your work as a designer is almost finished.

There are seemingly endless amounts of vases and structures to hold flowers. In fact, sometimes it can feel like there are more vases than there are actual varieties of flowers. To a certain extent, each designer must have their own personal philosophy when it comes to vases. When sending gifts, some prefer an expensive vase paired with less expensive flowers, while others prefer a cheaper vase coupled with

Restaurateur Maggie Hollingsworth is a partner at Blackfoot Hospitality, which owns several of New York City's hottest restaurants, including Little Owl, Market Table, and The Clam. Maggie is now doing flowers for the events her company hosts, and FlowerSchool was there to help her get a handle on the basics.

Choosing a vase that will work well with your flowers is an important step in the flower arranging process. To help your blooms come alive, you want to choose a vase that will make your flowers look like they're bursting from their vase, almost too beautiful to be contained.

elaborate flowers. As you will see in this book, I prefer to mix reasonably priced versions of both.

One thing's for sure: Your vase should have enough room for all your flowers and it should hold enough water for the flowers to drink. Furthermore, your vase shouldn't dwarf the flowers you want it to hold. In fact, using a vase that is one size smaller than you envisioned often gives the impression that the flowers are spilling out of the vase and makes them look more abundant.

More specifically, your vase should present the flowers you are using to their fullest potential, with the correct angles, an appropriate color, and the correct size. You should always imagine your flowers bursting out of their vase, as if the vase can barely contain so much beauty. When a vase is too small, it is pretty easy to spot. Not all the flowers fit. But when the vase is too big, it's a bit more of a challenge. It will appear as though the vase is swallowing up the flowers like Moby Dick engulfing a small whaling boat. Don't let this happen to you. Once the flowers have been prepped, you can follow the steps on page 112 to determine what vase you will need.

Bruce Littlefield, author of *Garage Sale America*, Weighs In on Vase Selection

The only thing that makes me happier than going out into my garden and cutting flowers is coming back inside and having a potpourri of vases to arrange them in. In fact, at every garage sale, flea market, junk shop, secondhand store, and auction I've ever gone to, I have always been able to find a vase, a bottle, an old oil can, a vintage pitcher, or an oddly shaped thingamajig that I can imagine a stunning collection of flowers wanting to be in. The joy for me is multilayered. Yes, the acquisition is a thrill, the making of the perfect arrangement is a creative pleasure, but the delight for me is often in giving it away. There's nothing like taking a gorgeous arrangement in the perfect vessel to a friend. Otherwise, at this point, I would have hundreds of such objects spilling from every shelf and cabinet.

In my collection, I have a varied selection in all shapes, sizes, colors, and materials. The one main requirement is that it holds water! The truth is, I mostly like inexpensive vases because I believe the flowers are the star, but over the years I have acquired, through garage sale good luck, tipsy shopping, and sentimental inheritance, several vessels that I protect like a mother bird because they are pricey, sentimental, or have an unbelievable backstory. Generally, if I'm out and spot a vessel that's a triple threat—eye-catching, unique, and affordable—I grab it. My flowers and I are always happy with such a purchase.

Four Tips for Finding the Correct Vase for Your Flowers

1. First, determine how big you want to make your flower arrangement and how much space you want it to occupy on the table. Professional florists typically use their hands as a guide, similar to the way people talk about a fish they caught or the size of a piece of cake they ate. If you are making something tall for a bar, then think tall. If you are making a centerpiece for a table setting, then think low and lush. Usually, I like the flowers to be three to four times the width of the vase. The flowers should also be two to three times as wide as the vase is high.

2. If this more precise means of determining the size of your arrangement isn't working for you, then pick up your materials in one hand. Your flower arrangement will be about as large as all of the materials held in your hand, which means the opening of your vase needs to be big enough to hold all of the material in your hand. This way of measuring isn't really mathematical, it's based on a florist's intuition and how much product you have.

3. As a rule of thumb, you should look at the opening of the vase to determine how many flowers will fit. Vases that have a small opening are good for taller flowers, and vases with wider openings are better for bushy flowers.

4. Visualize the faces of your flowers and how they will look best. Tulips look amazing upright, at attention, and with their stringing stems exposed. Hydrangeas (and garden roses, shown here) look good clustered together, often packed tightly, into a design.

These flower choices for vases are a guide and serve to give you a chance to feel the difference when determining how you want your flowers to look. I will make some suggestions based on how students have tried to put recipes together in the past and have failed. That isn't to say that those combinations are damned. Just that they can be challenging, though not impossible.

Common Shapes

Cylinder Vase

The cylinder is the first vase that florists should master. A cylinder vase can easily create a nice round shape, and will help new florists build an intuition for choosing the correct flower size. These are also ideal vases to have in different sizes as they can hold conditioned flowers that are waiting to be arranged. Generally, cylinder vases come in a standard size of 4 × 4 or 5 × 5. These vases usually require a minimum of 20 flowers, and they pair best with roses, hydrangeas, ranunculus, and other flowers with larger blooms.

PRO TIP: Add a candle for extra effect.

PRO TIP: Stack vases on top of vases for a special presentation.

V-Shaped Vases and Urns

There are several popular vase styles that share the same general construction: the V. This shape forces the flowers to start from a different point, and will not hold them in the same way. Urns are close cousins to V-shaped vases because of their similar shapes, and they should be approached in the same manner. This is a great vase to use with flowers that drape, such as spirea, euphorbia, and other long tendrilly things. This style of vase is more limited and is most appropriate for a garden-style bouquet that doesn't require you to see over it, like a table centerpiece. Flat-faced flowers like roses and ranunculus can pose a challenge here. Better to use items that have many flowers along the stem like spirea, delphinium, stemmed orchids, and long-stemmed tulips.

PRO TIP: V-shaped or conical vases lend themselves to materials with specific profiles. The distance from the center of a V-shaped vase to its edge is longer than you'd think.

PRO TIP: When using flowers with really heavy heads, you will need to be sure they are counterbalanced or else they can easily tip over. It is best to fill the outside edge of the vase and then fill the center so the vase won't tip over.

Compotes and Footed Vases

A "bowl" refers to two very different things depending on whether you're coming from the world of floristry or the world of dinner party planning. A compote is a dessert originating from medieval Europe that is made of whole pieces of fruit cooked in syrup. While everyone loves a good compote, in the world of flowers, the term "compote" refers more accurately to a footed bowl. The compote is best suited for a garden party or as a centerpiece. There are many reasons for this. First, compotes can be an amazing way to feature flowers as the centerpiece of a large table. It is nice when you can see a wild field of flowers across a table. Second, a compote doesn't hold a lot of water for the flowers to drink, so using them for a gift, delivery order, or weekly arrangement doesn't make any sense. Third, it is difficult to get lasting height out of flowers because of the shape of the well.

PRO TIP: Most flowers will need some kind of intervention in order to stay upright in a compote-shaped vase.

PRO TIP: A mix of fruit and flowers can work well if you're struggling to fill an entire compote with just the flowers you've got on hand. In a pinch, you can also simply fill the vase with some pears or other fruit.

Tall, Rectangular Vases

Tall, rectangular-shaped vases are some of my favorites to use both professionally and around the house. My favorite aspect of these designs is that they don't require a lot of flowers to look impressive. When you only have five of your favorite stems, this is the vase to choose.

PRO TIP: When the angle is more than 45 degrees, it becomes a challenge to keep the flowers standing in the vase. That is why many vases are just a touch taller than they are wide.

Square-Shaped Vases

These vases are a marriage between the compote-shaped vase and a cylinder vase. Flowers arranged in a square-shaped vase rest anywhere from a 45-degree angle up to a 60-degree angle. You will need an intervention of some kind in order to keep your flowers standing in the vase, at least until your skills for crisscrossing are good enough to make them stay on their own.

Bud Vases

Everyone should have a selection of bud vases, in a variety of sizes, shapes, and colors, in their collection. Bud vases are probably the most versatile vases and the easiest to get. You can mix a bud vase into any table setting, using any flower. Unfortunately, bud vases are also the hardest to clean. When buying a bud vase, you should look for the following things:

1. The hole in the vase needs to be big enough to fit the stem, or stems, you plan to put in it. I have seen many a flower jammed into the wrong size bud vase.

2. You should have many different heights and shapes within every color or style in your collection, so you won't have any trouble finding the vase that's right for your specific arrangement.

3. A bud vase should have a fairly large well for water so it won't tip over easily.

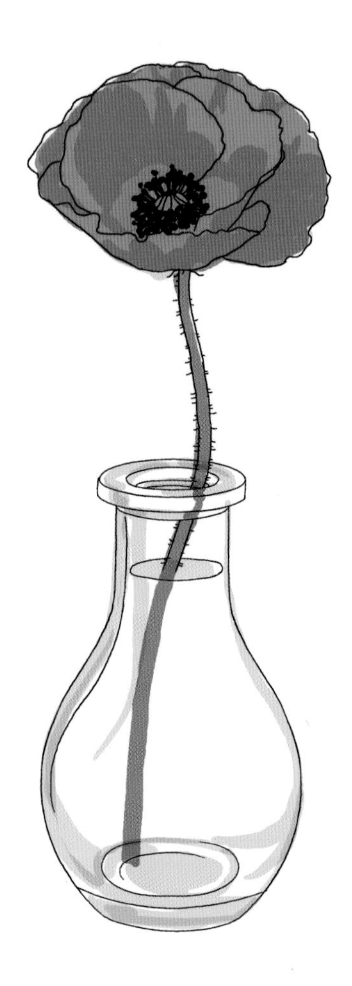

ABOVE AND RIGHT: *Students at FlowerSchool NY and FlowerSchool LA work on a variety of different stylish bouquets, creating everything from large statement arrangements suitable for grand hallways, to smaller, more intimate arrangements meant for home or the office.*

I cannot stress enough how important it is to make sure you select the correct vase before beginning your arrangement. Choosing the wrong vase is one of the most common mistakes that beginning designers make. If you want a flowing vase of flowers with lots of movement, you will be better off using a V-shaped vase instead of a tall vase with a small hole. The second most common mistake that I see is cutting all the stems too short right away. Inexperienced florists tend to cut the stems too short before realizing that it's impossible to make them longer once they're cut. The last crucial step is something we usually call the "florist's intuition." This means making sure you have chosen enough of the correct flowers, knowing what colors go together, having a wide variety of options without having too much variety, and knowing the style of flower arrangement you want to make before you actually purchase the flowers. These are all crucial tasks that you can learn over time, and these pages will serve as a guide to get you headed in the right direction.

A Basic Guide for Making Any Vase Arrangement

Start by inserting a quarter of your total flowers (not including foliage) into your vase, turning the vase as you add each individual stem. Be sure to keep your star flowers—any flowers you want to display prominently or any especially delicate flowers—for the very end.

Make sure that you keep your stems long enough to touch the bottom of the vase. This will give you the flexibility to change the overall size and shape of your arrangement as you see fit.

Make little cuts (no more than a quarter inch) so that you don't cut your stems too short. You can't make them longer easily.

Once you've placed the first quarter of your flowers, step back and take a look. Ask yourself: Are the stems too short or too long? At this point, the answer should be too long.

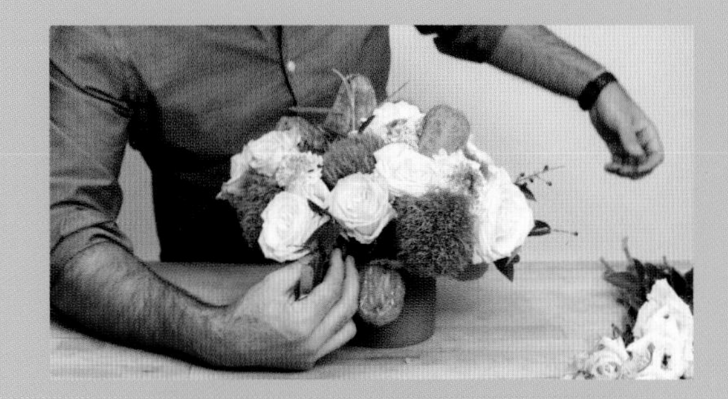

Now add a quarter of your foliage materials (if you're using them) and the next quarter of your flowers. Again, turn the vase after you add each new stem to ensure that you're evenly distributing your materials. Once you're done, step back for another look. Ask yourself if the stems are too short or too long.

Finally, add the last half of your materials. This will include all of your most delicate flowers and any flowers you want to display prominently. At this point, all of your materials should be inserted into the vase, and you can trim any stems that are too long.

PRO TIP: When working with flowers, it is very important to only handle them by their stems. *Never* squeeze or push the flowers into place—you will only damage them.

CYLINDER VASES: Above is a small selection of different cylinder vases to serve as inspiration for you when you shop for your own vessels. This classic shape pairs perfectly with flowers that feature larger blooms, or with flowers that are side-facing, like sunflowers or dahlias.

V-SHAPED VASES: On this page you'll see a few different examples of V-shaped vases to help inspire you. These shapes go best with longer flowers that tend to drape, and need a little extra support, like spirea, droopy garden roses, branches, and other oddly shaped items.

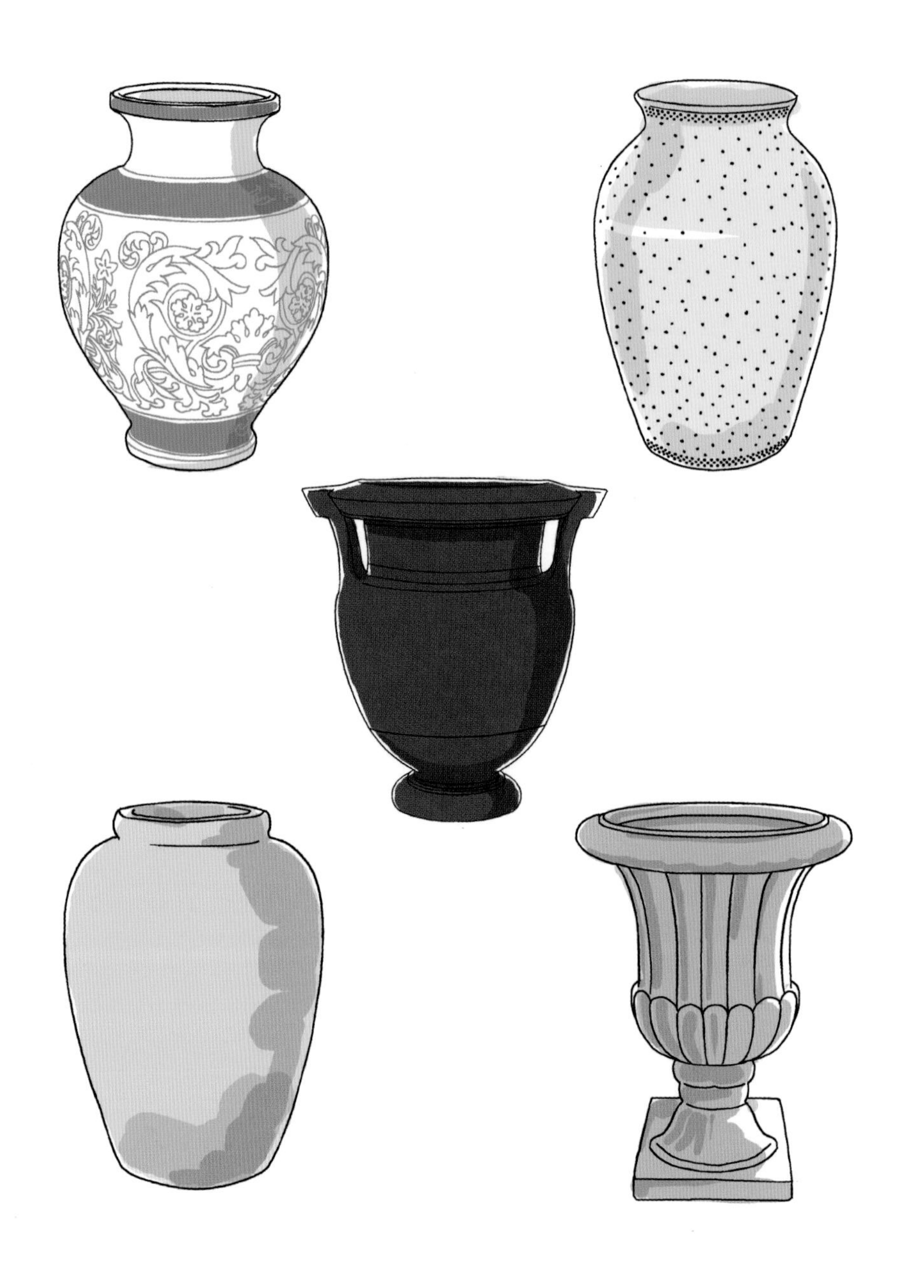

URN VASES: Urns are perfect for creating garden-style, statement arrangements that will stand out from the crowd and draw the eye.

COMPOTE VASES: A compote or footed urn is the ideal shape for creating arrangements meant for dining tables and entryways. These shapes offer great draping capabilities, but they won't impede your eyeline as they tend to be fairly short.

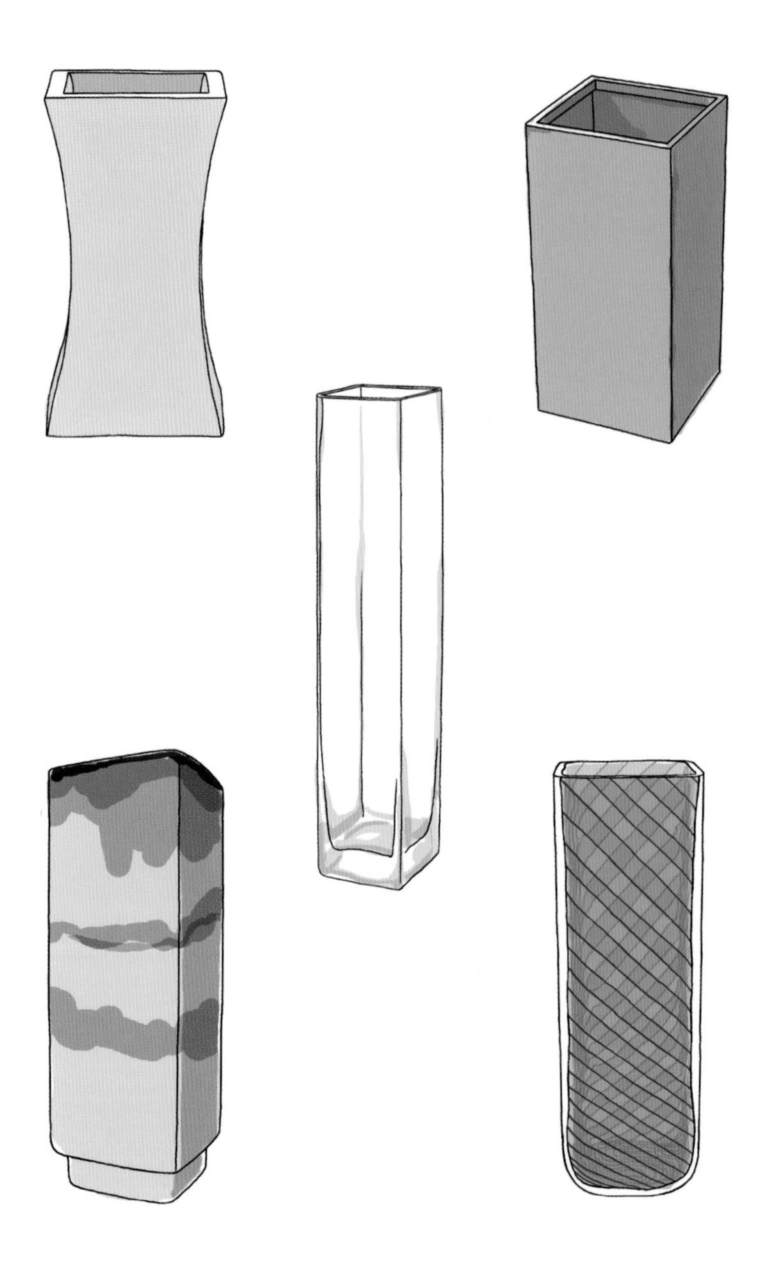

RECTANGULAR VASES: This classic shape is a jack of all trades and is perfect if you're running low on materials. When you only have a bloom or two on hand, a rectangular vase should be your go-to selection.

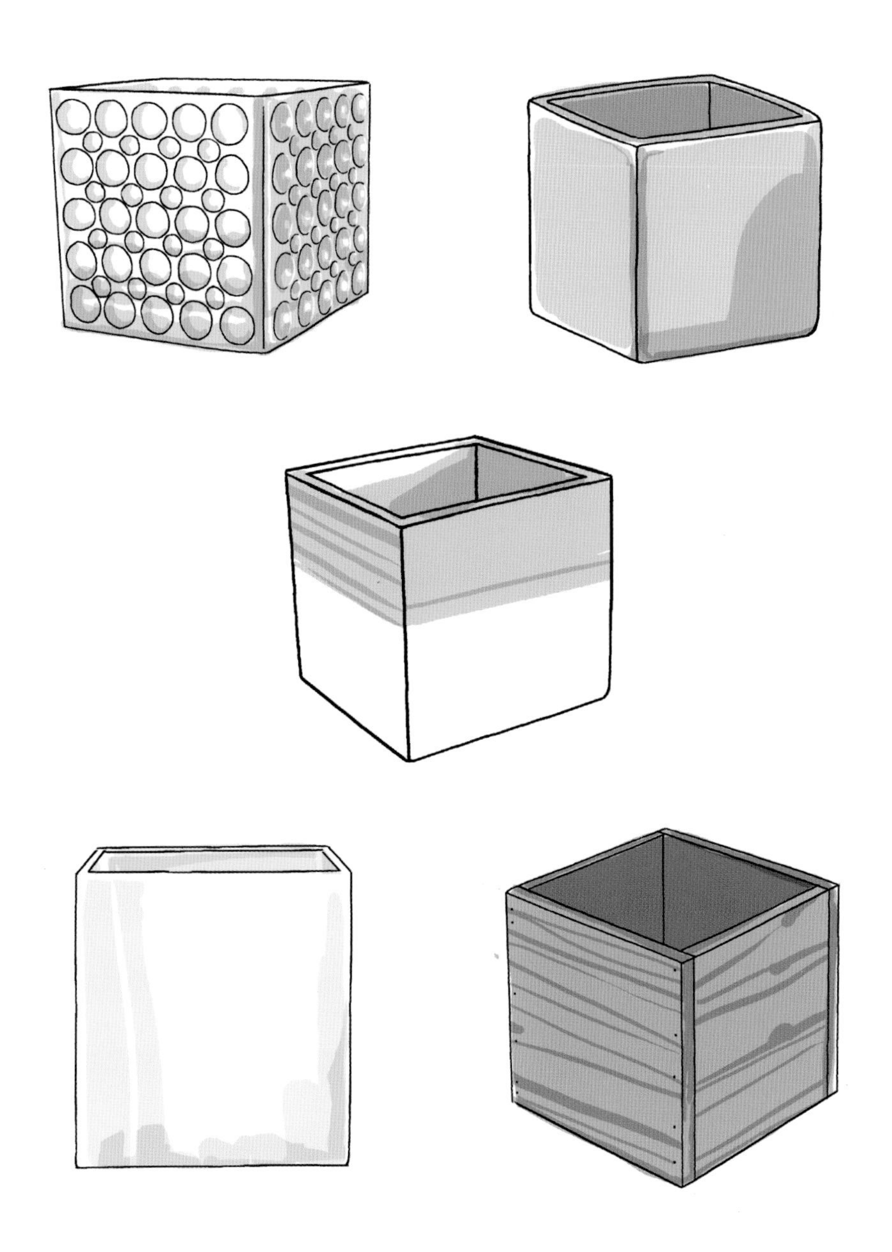

SQUARE-SHAPED VASES: A square-shaped vase is the ideal vessel for anyone hoping to create a sleek, modern arrangement.

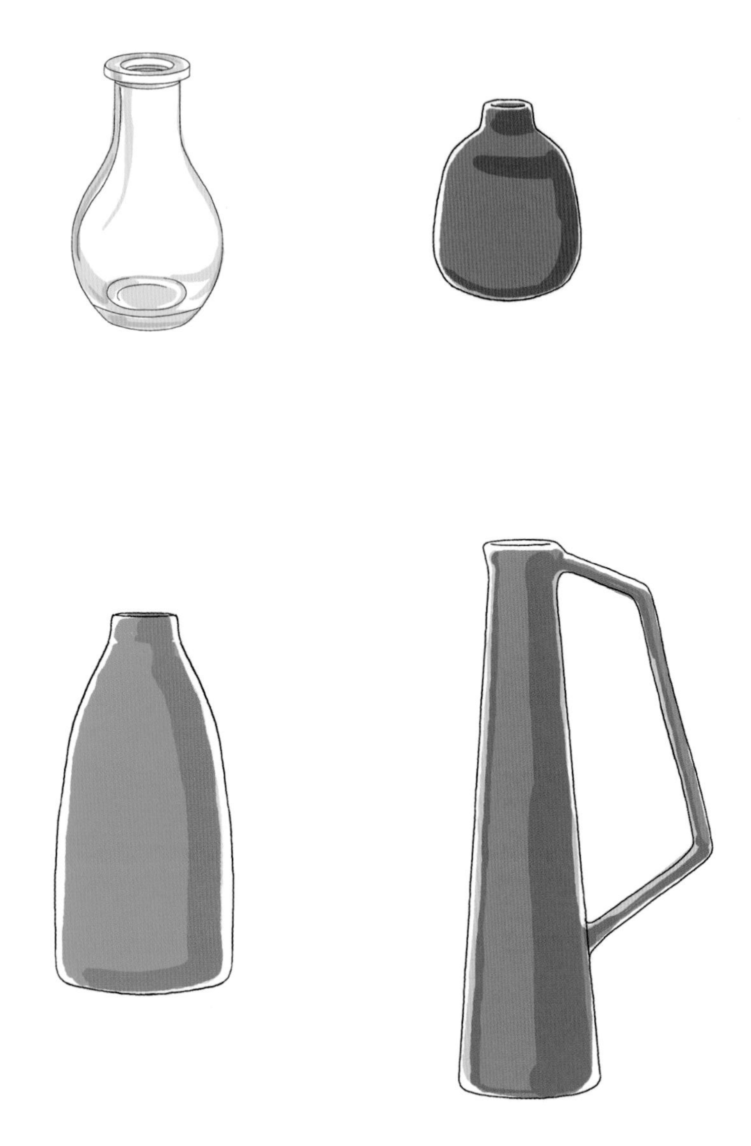

BUD VASES: A bud vase can be tall or short and it can feature a wide-mouthed opening or possibly a small-mouthed opening. These wide-ranging vases are perfect for displaying a single bloom to its best effect.

5.

Working with Flowers

LEFT: *Flowers from your garden are always ideal for arranging, but make sure you hydrate them properly to ensure they last as long as possible.* **ABOVE:** *A completed arrangement is displayed on a worktable at FlowerSchool New York.*

WHEN PURCHASING FLOWERS FROM A FLOWER MARKET, CUTTING FROM YOUR GARDEN, OR BUYING FROM YOUR FAVORITE GROCERY STORE, IT IS OF PARAMOUNT IMPORTANCE TO GET THEM HYDRATED OR REHYDRATED AS QUICKLY AS POSSIBLE. In the flower business, we use the term "condition" to cover a host of important preparatory steps you must take after you've purchased your flowers but before you begin to arrange them. Here is a basic, step-by-step process to follow each time you bring new flowers home:

Step 1: Cut Your Flowers

I'm a firm believer in using a floral knife to make flower arrangements. Using a floral knife is faster, more efficient, and better for the flowers. There are several steps that florists must learn in order to work safely and efficiently and to ensure that their flowers last as long as possible. These steps make up the very foundation of what we do and they are outlined below. All flowers should be freshly cut as soon as you arrive at your workstation. Lay all of your flowers down on the table in front of you and be sure to have a garbage can close at hand to catch any cuttings. Flowers should be cut on an angle so that their stems do not rest flat on the bottom of the vase or bucket you've chosen for storing your flowers. Cutting on an angle provides more surface area, which lets the flowers drink in more water.

HOW TO CUT

1. Hold the knife in your hand between your knuckles and close your fingers around the knife handle.

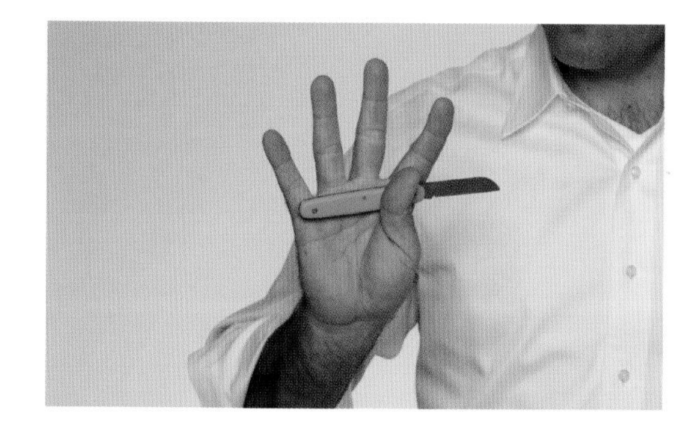

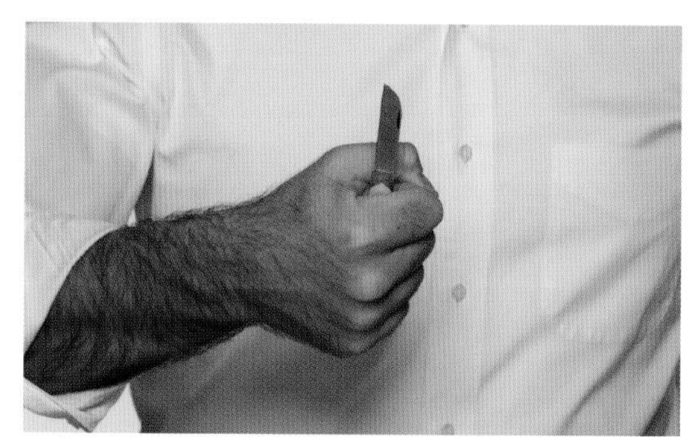

2. Next, hold your thumb and pointer finger in parallel. Move the flower to the knife, not the knife to the flower.

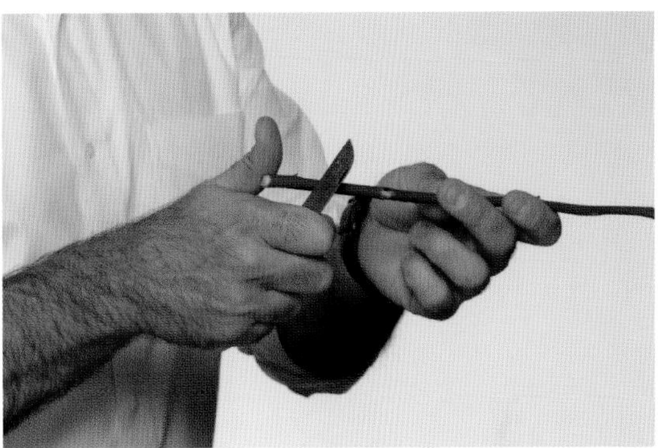

3. Dig into the stem slightly and then pull the flower away.

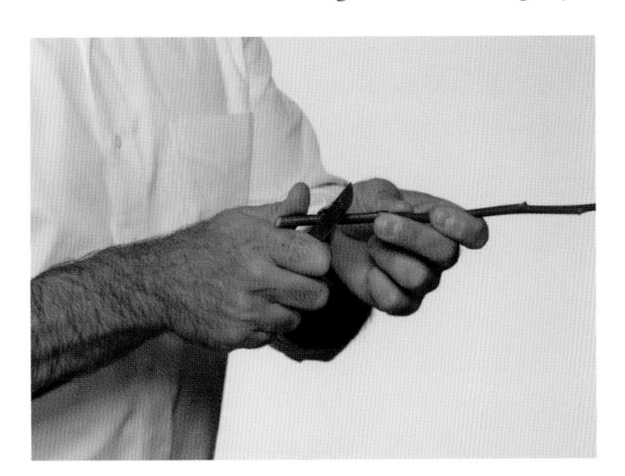

PRO TIP: It is best to start out making some practice cuts on flowers, twigs, or branches that you're not going to miss if you make a mistake.

4. Once you've become proficient in this basic cut, you are ready to work on making shorter cuts. To do this, use your hand and your knife to cut the stem in a pulling motion, using your thumb as a guide and keeping it slightly close to the blade.

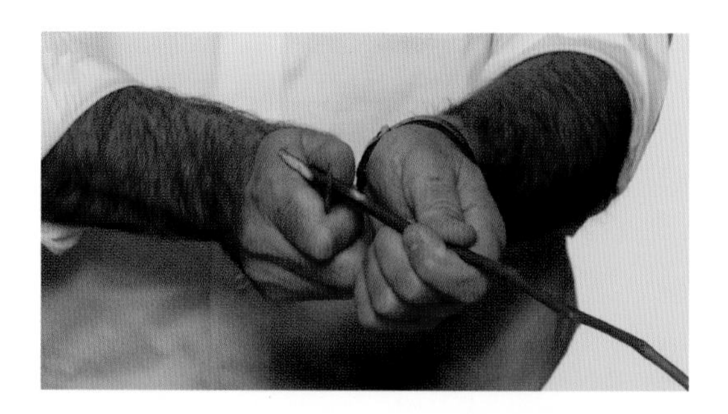

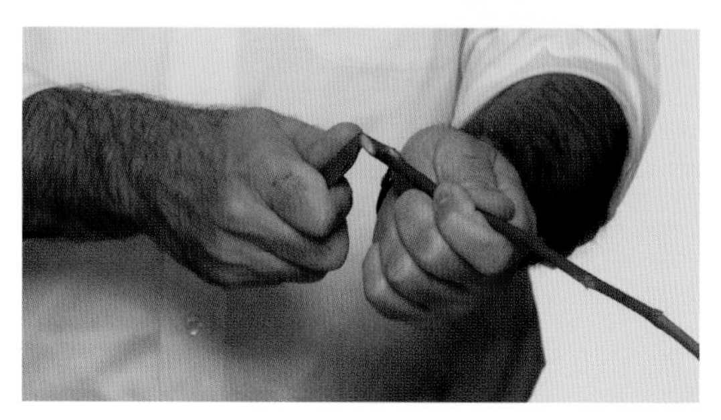

5. Once you gain more confidence, you will be able to cut multiple stems at once. To practice, create a "pave," or pile, of organized stems on a table, making sure they're lined up and not crossed. Pick up the pile in one hand and, using your knife skills, give all of the stems a cut on a slight angle. You will see how much faster it is.

Step 2: Remove Any Foliage

In order to get many varieties to drink more, it is best to remove most or all of their foliage. Not everyone follows this rule because the foliage that comes with a flower can often be quite attractive. I myself have been known to keep a leaf, or two . . . or more. If longevity is what you seek from your flowers, then you should remove all excess foliage from each stem. Remember to purchase some stand-alone foliage (such as *Ruscus*) to mimic the foliage that you have removed. The entire arrangement will last longer.

To remove foliage, you can use the "strip method," running your fingers down the stem to remove all the foliage in one pull. If you prefer, you can use your knife to do this, which can be helpful if there are thorns or dangerous barbs on your stem. If you do opt to use a knife, be careful not to cut into your stem.

PRO TIP: Flower stems are genetically trained to heal when cut—they clot and heal themselves, similar to cuts on human skin. So make a cut quickly and get each stem into water immediately.

Step 3: Prepare Your Water

When flowers are bought fresh from a farm or grocery store, they have typically been out of water for a long stretch of time. The ends of the flowers' stems are dry and, in some cases, there will be air pockets in the stems that prevent the flowers from getting a good drink. Giving your flowers clean water that is the right temperature will do wonders for their hydration. Make sure you condition your flowers for hydration in an organized way, spiraling all your flowers into a vase and then arranging them in your vase. This will give you ideas on how each flower will behave, and you may even learn from this process, which is usually thought of as busy work. In fact, in my former shop, I have had customers simply purchase my hydrating flowers right off the shelf as if it was a complete arrangement! Conditioning your flowers will help you see the ways that your flowers naturally look and feel in a vase.

WHEN TO USE WARM WATER

Flowers that bloom in warm climates or during summer months should be given a bath of warm to very warm water with plenty of an antibacterial solution because, as discussed, bacteria grow faster in warm water. Flowers in this category include, but are not limited to, roses, hydrangeas, and hypericum berries.

WHEN TO USE COLD WATER

Flowers that bloom in the spring and the fall tend to do better when cut with a knife and hydrated in cool water with an antibacterial solution. These flowers include tulips, hyacinths, daffodils, and dahlias.

HOW TO "CLEAN" YOUR WATER

Water needs to be clean and free from bacteria. If you live in an area with poor water, you may benefit from either using filtered water or adding extra flower food to your tap water.

> **PRO TIP:** If you're not sure whether to use warm or cold water, look at the stem anatomy of whatever flower or foliage you're working with. Branchlike materials such as hypericums or actual branches require warm to hot water, while thin, soft-stemmed flowers like calla lilies require cool water so their stems won't melt.
>
> **PRO TIP:** Calla lilies will rot under water, so it is best to use only 1–2 inches of water in the vase for longevity during the conditioning process.

To Rest or Not to Rest

All flowers should have some time to settle down in a vase of familiar clean water before you attempt to use them in an arrangement. There is some latitude, however. Here is a quick cheat sheet of flowers that need more or less time than usual:

ANEMONES: Ready right away
FOLIAGE: Ready right away
TULIPS: Need 4–6 hours hydrated, and will take 3–4 days before they fully open
AMARYLLIS: Need a full 4–5 days before they'll open
LILIES: Can take up to 7 days to fully open

Step 4: Add Your Flowers and Clean Up

Now that your flowers are ready, and your water is clean, all that's left to do is add your flowers to the water. Once you're done, make sure to clean any tools you used so that bacteria does not build up on the blades.

PRO TIP: Tulips and other flowers that need support to keep them from drooping should be hydrated in a cone in order to stay upright during the process.

PRO TIP: Hydrangea and lilac stems can be smashed with a hammer and wrapped with a wet paper towel to keep them hydrated.

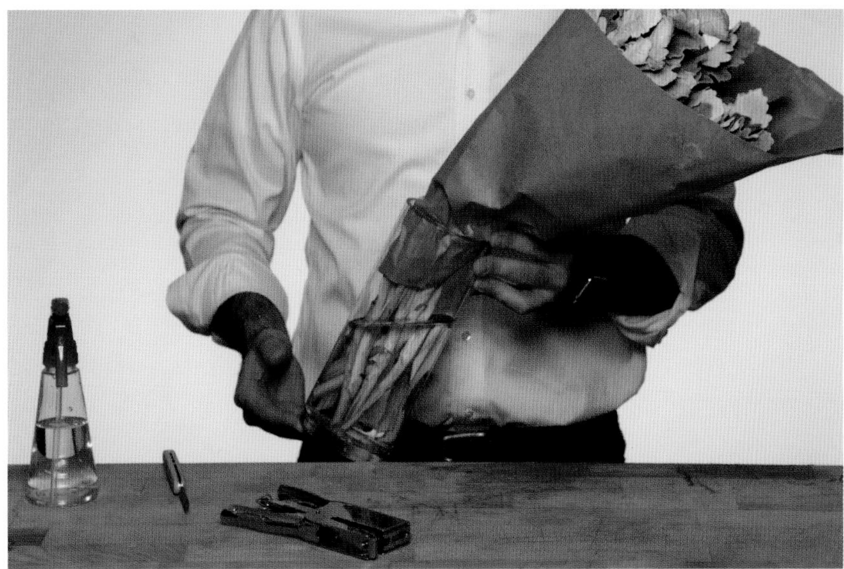

If you have flowers or foliage with longer stems, we always recommend that you create a cone for your materials. This will help them maintain their shape and will ensure that they don't droop. This is a very helpful process for tulips, dusty miller (as shown above), and other flowers that tend to flop over. For extra hydration, spray your materials with water before coning.

Special Tips for Working with Flowers from Your Garden

I often hear stories from people who cut from their gardens and the flowers poop out straight away. There are several factors that can cause a homegrown flower to die an early death. I believe that these steps should mitigate most issues.

1. **USE CLEAN WATER:** Gardens are laden with bacteria. Blooms cut from the garden can benefit greatly from a hydrating, bacteria-killing solution.

2. **CUT EARLY IN THE MORNING:** Be sure to cut whatever you need very early in the morning before the plant wakes up. Otherwise, the stem will think that it has been killed.

3. **LEAVE THEM ALONE:** Once you've cut and added your flowers to clean water, leave the flowers alone to hydrate for 2–4 hours before attempting to use them in an arrangement.

After all these steps, sometimes it can seem easier to just call your favorite florist and let them do all the work. But trust me, once you get these steps down, you'll be able to create any arrangement you like, rather than being at the mercy of whatever is for sale at your local florist.

Doing flowers is about accepting life's moments as they're thrown at you. Just as in life, flower arranging is all about playing the hand you're dealt. You may have a plan to create a beautiful arrangement of peonies only to find that there are none to be had at your local store. If you've taken the time to master the skills we've discussed, you'll be able to use your knowledge to think on your feet and find a stylish solution. Having really good style, and making beautiful arrangements, is not about

It's important to have a plan before you begin to shop for your flowers, but you should also be prepared for your plan to change. Maybe the flowers you want aren't available, or maybe they don't look as wonderful as you'd hoped. If this happens, take a breath and then see what other flowers are available that work within your chosen color story and style.

what materials you use. If you've mastered the techniques in this book, you should be able to make any group of flowers look great. As long as the foundation is there, you'll always have an arrangement that looks gorgeous, no matter what materials you choose. As you begin to shop for your flowers, you're likely to come across a wide variety of blooms. You will have trouble putting Eremurus together with roses, but not tall lilies. Roses, poppies, and anemones should go together easily. Use these pages to help determine a flower recipe when purchasing for your home creations. These illustrations will be your guide to making sense of flowers by shape, size, and profile.

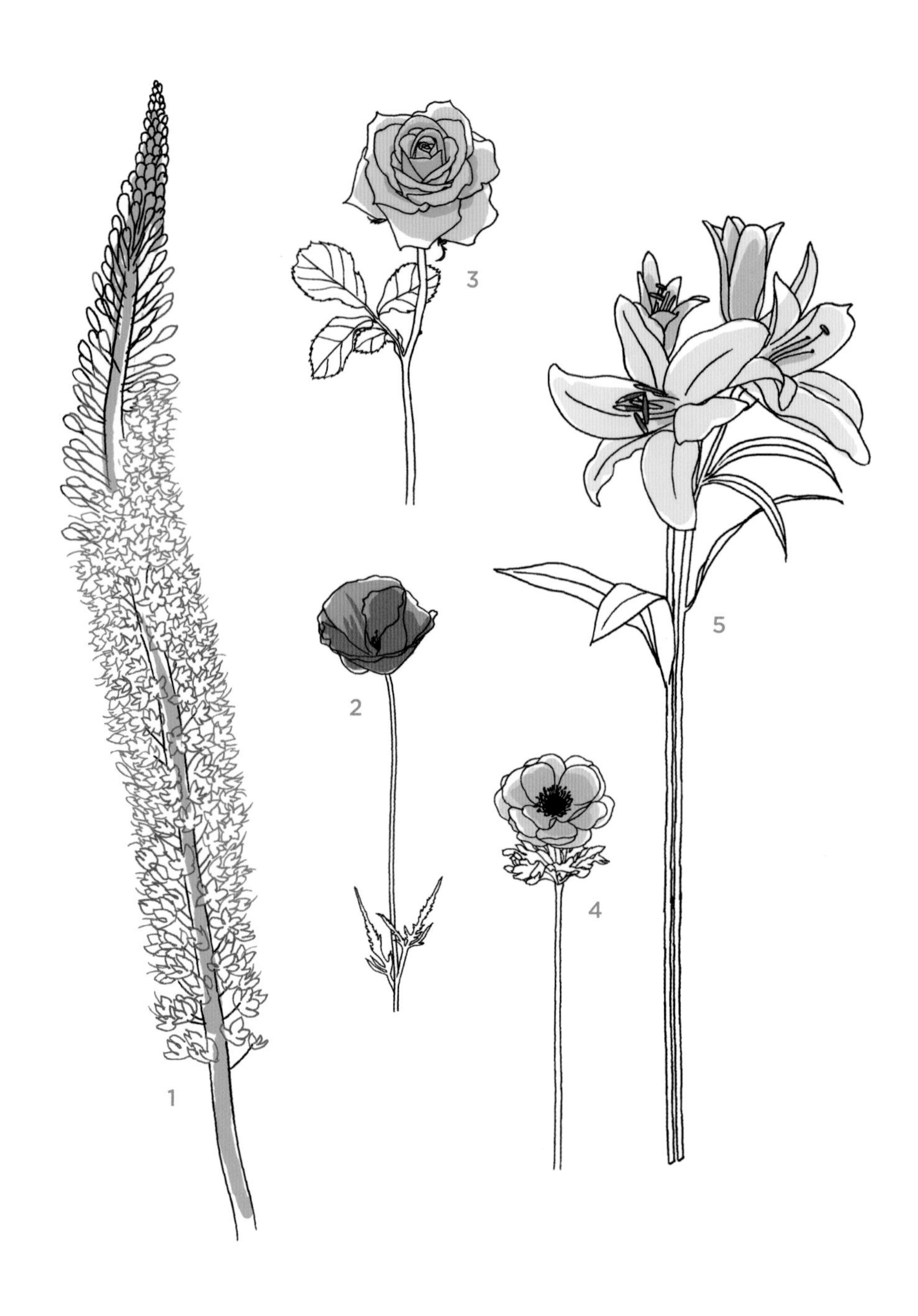

1. Eremurus 2. Poppy 3. Rose
4. Anemone 5. Lily

1. Ranunculus 2. Orchid 3. Hydrangea
4. French Tulip 5. Carnation 6. Phlox

1. Lilac 2. Amaryllis 3. Calla Lily
4. Tulip 5. Delphinium

1. Peony 2. Spirea 3. Dahlia
4. Green Trick Carnation 5. Monkshood

6.

Techniques for Arranging: Common Styles and How to Make Them

CHOOSING THE STYLE OF FLOWER ARRANGEMENT YOU WANT TO MAKE SHOULD BE YOUR FIRST STEP ANYTIME YOU SET OUT TO CREATE A NEW DESIGN. Once you've chosen the style (rustic, modern, garden, etc.), it will be much easier to choose the flowers you will need and then the vase that you will need. Before determining what style flower arrangement you want to build, you should start with an understanding of the different basic styles (see Chapter 3).

In this section, we will touch on some basic elements of style to help guide you on your journey. This, hopefully, will take much of the guesswork out of the process of becoming a florist.

As we have been establishing in these pages, your working order of operations should be:

1. Determine the colors and style of flower arrangement that are appropriate for your needs. Make sure you take into account the vase you'll be using when determining color and style.

2. Gather your organic materials from the grocery store, your backyard, or in the great wild.

3. Condition your flowers, giving them a fresh cut and trim and storing them at the correct water temperature until they're ready to be used.

4. Make sure you have your vases ready, and that they're the correct size and shape for your flowers.

5. Build your arrangement.

6. Repeat these steps each time you begin a new arrangement.

A student at FlowerSchool New York begins to build a new arrangement during a Master Class taught by Lewis Miller.

Basic Techniques for Making All Flower Arrangements

Based on the order of operations we've laid out on the previous page, this section will focus on Step 5: build your arrangement. We will illustrate how to "see a bunch of flowers," think of a design, and then effortlessly re-create that design. When working within a style, you will find yourself liberated from the notion of having to work with specific flowers and you will become more focused on the whole design. My favorite story about choosing flowers comes from a flower market buying trip I once took with the great Christian Tortu. "Cal, I would like this," he said, pointing to a very expensive imported flower. I replied, "Christian, that is too expensive." Then he said, "Okay, then this." And he chose another flower. It was in this moment that I realized that it was not so much *what* flowers Christian was using, but rather *how* he used the flowers that were available to him.

If you choose a certain style and follow the framework laid out on the next pages to create this style, you will always end up with a beautiful bouquet, no matter what flowers you choose. Knowing how to build your style without tying it to a specific flower is key. Style is all about color, flower profile, and flower size. The specific type of flower you choose is of secondary importance. Keep this in mind the next time you head to the florist to buy lilies (or daisies or delphinium) only to find that there are none available. Knowing what style of arrangement you want to make will give you the freedom to make last-minute changes in your materials, if needed. With this in mind, let's first run through the different design styles.

Master Florist Takaya Sato of L'Atelier Rouge teaches a class at FlowerSchool New York.

Classic Spring Modern

When we're thinking of spring, the first thing we should consider is color. In this demonstration, we are using yellow as our focal color with cream to help soften it. Once our color story is set, we need to determine which vase to use. In this case, I've opted for a brown container. It's organic in nature, which will accent the deep yellow flowers to create a rustic yet modern look.

1. First, choose a container, which can be anything from an old tin can (or a new one with a bit of age) to a mason jar to a vase wrapped in birch bark. If nothing is available, a simple brown or cream container will do.

2. Once you've chosen your vessel, you'll need to decide what flowers you want to use. You should either cut from your garden or purchase flowers in colors that are most closely associated with a walk in spring. What do you typically see on a springtime walk? Usually brightly colored daffodils and forsythia branches.

3. Be sure to have at least 7 to 10 different varieties of flower, and 3 to 5 stems of each.

4. For a wild look, make sure that the flowers are not placed too close together. Nature never smashes flowers and leaves together. They are given just enough space to live but not too much to not be part of a unit. To create a more modern look, place all the flowers close together in order to build a strong shape. In the following images we're creating a modern look.

PRO TIP: For any arrangement, you should work in quarters. Insert a quarter of your flowers and materials and then step back and turn your vase 45 degrees. Then, insert the next quarter of your flowers and step back again to turn your vase. Keep turning and inserting until you have placed all of your flowers.

STEP 1: Once you've filled your cylinder vase with water, place your hands around all of the flowers you plan to insert into your vase. This will help you visualize how wide your flower arrangement should be.

STEP 2: Taking into consideration the size of the flowers you plan to use, extend your hands out from the sides of the top of the vase and stop when you reach a distance that feels right to you, and that is in proportion to the size of your vase. This distance represents how far you want your flowers to extend beyond the lip of your vase.

STEP 3: As mentioned previously, do not start by jamming all of your flowers in all at once. Instead, add a quarter of your flowers around the outside edge of the vase. Make sure you hold back your most delicate and most beautiful flowers to insert last. When using materials such as forsythia or other wild and branchy items, be sure to separate them into smaller, more usable pieces. Larger pieces may seem like they are working for you, but later you will find that they are too big and can't be adjusted easily.

STEP 4: This step is VERY IMPORTANT. It's time to start cutting your flowers to fit your vase, but make sure you start cutting conservatively! You can always cut again to make your stem shorter, but you can't add stem back if you cut too short. Proceed with caution and take your time.

STEP 5: Now it's time to insert a quarter of your extra greenery, making sure that whatever foliage you choose is a bit shorter than your flowers, which should be the star of the show. In this case, we are using flowering branches of forsythia as our foliage. When you're done with this step, half of all of your material should be in your vase. Take a step back and look at what you've done so far. Is it balanced?

STEP 6: Slowly begin inserting the rest of your flowers and foliage, but take care to reserve two or three pieces of foliage for final touch-ups. After each insertion, rotate your vase slightly and then insert another flower or piece of foliage. By constantly rotating your vase, you'll make sure that all of your materials are evenly distributed and that your final arrangement is symmetrical. If any of your stems are too long, now is the time to make your final cuts. As always, move slowly and deliberately.

STEP 7: Once you've added all of your base materials, you're in the home stretch! Step back from your arrangement once again, and check to see where the other supporting flowers and colors fit.

STEP 8: It is now time to add your other colors. Here we are adding green to help soften the bright yellow and ground the arrangement in the natural world. However, you may prefer to stick to just yellow. This design choice is yours to make.

Now it's time to enjoy the fruit of your labors! This arrangement is perfect for your desk at work, or in your home office. It would also make a lovely centerpiece for a table in your front hallway. Or, better yet, why not gift it to a friend or loved one?

Ingrid Carozzi: The Student Becomes the Master

Ingrid Carozzi is an alumnus of FlowerSchool who has gone on to become one of the most celebrated designers working today and a Master Florist in her own right. She is the owner of Tin Can Studios in Brooklyn, New York. In addition to providing full-service floral design for A-list events and collaborating with major organizations like Facebook, LVMH, and the New Museum, Ingrid is also the author of two celebrated books: *Brooklyn Flowers* and *Handpicked*.

Ingrid makes arrangements that pair gorgeous blooms with recycled or reused vessels, creating a stylish and much-celebrated juxtaposition of old and new. Luckily for us, she's taken a few minutes out of her busy schedule to offer some insight into her process. One look at the gorgeous arrangements on the following pages, and I'm certain you'll be just as big a fan of Ingrid's as I am.

What makes your style unique?

Finding the right materials. No matter where you are in the world, you can look for things that are outside of the norm, and that will take your arrangement from ordinary to extraordinary. It's like being a chef who's hunting for exotic ingredients. Finding two or three interesting materials to include in your work can really elevate the final product.

What's your approach to color?

I like paying attention to color trends; they allow the flower arrangements to feel current without going overboard.

What makes a great florist?

A great florist is simply a hardworking person who is ready to roll up their sleeves. You have to know how to condition and handle flowers, but you also have to be humble and willing to take direction. Don't be afraid to ask questions and make sure you know what's expected of you.

What makes a great designer?

Art school is a great start. You can become a great designer without going to school, but you must have a good eye for color, form, and texture. Also, you've got to be able to work well with clients. If you are a designer, you are translating a message with flowers. That's your job, to deliver the message. When you're working with a brand, it is extra important that your flowers convey the message that the brand wants. Make sure you understand the vision of whomever you're working with or for. Making an arrangement without having a clear vision in mind is very difficult.

What advice would you give to a florist who is just starting out?

To start: Look and see what's out there! When I go to the market, I find one flower that speaks to me, and I build everything around that one flower. If you're paying attention, there will always be something that speaks to you. If you're making something for a special occasion, you may want to bring in some seasonal elements. It all depends on what this arrangement is for, and what your vision is.

A Lesson in Compote Arrangements with Ingrid Carozzi

On a recent afternoon in New York, Ingrid stopped by the FlowerSchool studio to offer us a lesson in the art of arranging flowers. For this lesson, she selected a compote vase, which requires chicken wire in order to hold the flowers in place. With her vase now set up, and her conditioned flowers right at hand, Ingrid is ready to give us a behind-the-scenes glimpse at her artistic process.

Ingrid begins her arrangement by placing about a quarter of her materials into her vase, building a border around the edges of the vase. Then, she makes her larger design come to life by putting in some longer-stemmed roses to build the overall shape.

Next, Ingrid builds the center of the arrangement, making sure to take height into account as she places her blooms around the compote vase. She uses tall flowers to extend outward, and very short flowers to fill space.

The time has come to add a little foliage for balance. As we've discussed previously, foliage is a key ingredient in any successful flower arrangement. If you're ever at a loss for what color to use, think green!

With a fair amount of her materials in the vase, Ingrid takes a moment to stop and check how far her blooms are extending over the side of the vase. Now is an opportune moment to make any adjustments to ensure the arrangement has the proper sense of height and proportion.

Here's a close-up image showing the finer details of Ingrid's arrangement. Notice how she is only handling each flower by its stem and she's giving each bloom space. She's also incorporating unique materials like berries in order to add another layer of texture, color, and surprise.

Time to take a moment and check her progress. As always, now is the time to consider the proportion of the arrangement and the distribution of materials and color. If everything feels balanced, it's on to the final step.

It's now time to add the final reserved pieces of flowers and foliage to the top of the arrangement.

Ingrid notes that her work in flowers is a lot like fashion. It is important to have two or three very high quality garments that go with your sensibility and style. The rest of your wardrobe should just augment those garments. If you pick a special or hard-to-get flower in a unique color, the rest of your arrangement should serve to support that flower and color.

Modern Style

My first thought about modern design is the Frank Lloyd Wright–designed Guggenheim Museum in New York City. It's all one color with clean, minimal lines and every part of the building serves a specific purpose. Thinking of the modern style also calls to mind the work of painters like Mark Rothko or sculptors like Isamu Noguchi. These artists created organic and human forms, but they realized these forms in the simplest way possible, and that's what we hope to achieve in the modern style of flower arranging. Beauty through simplicity.

1. The modern style can be achieved by using all one color, one shape, or one variety of flower. Ideally, the colors you choose will match.

2. Once you've chosen your materials, the next step is to think of a shape. The most common shape in the modern style is a dome, which is perfect for a coffee table arrangement or for an arrangement meant for a countertop or as a dinner table centerpiece. The best approach to this type of arrangement is a 1:1 ratio, meaning that the length of your flower stems should be the same height as the vase you choose. If you're hoping to create a more robust impression by using larger materials, try a very tall vase with a smaller opening (see page 80).

3. In the floral industry, this design is sometimes referred to as a "rush," as it comes together quickly because it doesn't always require a large amount of flowers, or different varieties of flowers.

STEP 1: For our modern arrangement, we've chosen one color to work with: pink, and we've opted for the domed shape that is a classic facet of the modern aesthetic. To begin, place four of your roses in a square-shaped vase, one flower in each corner, making sure that its stem touches the bottom corner on the opposite side of the vase.

STEP 2: Once you've placed your four flowers and trimmed them so that the blooms are extending over the sides of your vase at the preferred length, your next step is to begin inserting additional stems. Since a modern arrangement typically only includes one type of flower, in this example we're adding more roses.

STEP 3: As you begin to insert more of your roses, you'll want to create a circular border around the outside edge of your square vase. This will serve as the base of your dome shape.

STEP 4: Now that you've built the base of your dome, it's time to step back and check your work. Are your flowers extending too far over the edge of your vase? If so, now is the time to trim your stems.

STEP 5: Once you've confirmed that the work you've done so far is to your liking, it's time to start filling in your arrangement. Working from the outer edges, slowly begin to add additional blooms to the middle of your arrangement. Remember to make small cuts, and handle your flowers only by the stem, taking care not to damage any of the other flowers as you place each new bloom.

STEP 6: To complete the dome, you'll need to make the flowers in the center progressively shorter, or else the dome shape will never be achieved. This is an important time to move slowly, and only cut a stem when you're absolutely certain of the length you need. If you're unsure, place your stem next to your vase and use that as a visual cue before cutting your stem.

STEP 7: As you continue to build your dome, you'll want to make sure that you're adding an abundance of flower stems into the vase, so that all the stems are firmly packed in and holding each other in place. It's equally important for the stems to be propped up on the bottom of the vase to provide additional support, so don't cut them too short.

STEP 8: Time to step away and take one last look at your arrangement. Are there any empty spots? Are all of your blooms the correct height? Now is your last chance to make any changes.

Garden Compote Style

The term "garden style" can be confusing because there are many different types and varieties of gardens, ranging from a moss garden outside of Kyoto to the gardens at the Palace of Versailles. In contemporary floral design, the concept of garden style or garden design is most closely associated with an overgrown formal garden that contains both invasive weeds as well as designer flowers. It is this mix of shabby, overgrown chic with organic forms that entices the eye, evoking memories and nostalgia for days past. A garden-style arrangement should resemble a completely overgrown garden with flowers and foliage moving in every direction. Although this design should look wild and untamed, the style can be achieved by following these steps:

1. Picking the appropriate vase is of paramount importance for the garden style. When thinking shabby chic, you can't use a simple glass cylinder or anything that's too new looking. A moss-covered terracotta container or a faux mercury glass compote will do the trick (faux is preferred here as mercury can be toxic).

2. Choose four or five different types of foliage (see image in Step 1) that are different shapes. Think of a garden where the moss and boxwood have taken over and overgrown their welcome. These should not include fancy foliage, but rather authentic weeds.

3. Mix these flower bases with designer blooms fit for a royal garden. Imported garden roses, anemones, and clematis are industry favorites. You will need a minimum of seven different blooms. We recommend three designer bloom varieties and four bloom varieties to help support the overall aesthetic.

4. You should start your arrangement by focusing on the edges. This is when chicken wire and/or a tape grid (see images in Steps 2 and 3) can come in handy, depending on your vase selection. When making this arrangement, keep in mind that the flowers are meant to spill out of the vase in order to create an overgrown look (see images in Steps 4 to 6).

PRO TIP: Put your focal flower in at the end (see final image).

STEP 1: As ever, your first step with this arrangement is to place your hands on each side of your vase to get your mind oriented to how wide the final arrangement should be. It can also be helpful to lay out your flowers based on the color story you've chosen for your arrangement, as shown here.

STEP 2: Most compote arrangements require some kind of intervention in order to help your flowers stay in place. In this case, we're using chicken wire, which has been cut to fit our vase.

STEP 3: Once you've got your chicken wire cut and placed into the vase, use some of your floral tape to hold the chicken wire in place. Make sure to test this out with a stem or two to be certain that the flowers won't eject.

STEP 4: Now it's time to start inserting your flowers. As mentioned previously, start by placing a quarter of your flowers around the outside edge of the vase. This will give you an idea of how big the arrangement should be. It should also match up with the amount of product you have.

STEP 5: Next, insert a quarter of your greens, making sure that they are a bit shorter than your flowers, which are the focal point.

STEP 6: Using your remaining flowers, insert stems around the outside edge of the vase in a circular motion. Insert a flower, then turn your vase. Insert another flower on the opposite side, and turn. Keep going until the outside edge is completely filled. Finally, insert your remaining pieces of foliage into the top of the arrangement.

Contemporary Elevated Style

This style is often confused with modern style, but with further research you will find that the two styles have very little in common. Contemporary style can be wild in shape, incorporating a mix of all types of flowers and foliage, and typically will not be placed in a basic 5 × 5 vase (see Chapter 4 for a discussion of vases). This style of flowers more closely resembles art than the typical floral gift. Think of an installation rather than an anchor piece in a classic urn. Sometimes, contemporary arrangements don't include any flowers at all but rather showcase a collection of sticks artfully put together to form a piece of sculpture.

1. For all practical purposes, I think contemporary style is a fun way to arrange flowers, and it can lead to creating some unexpected yet beautiful results. This design is usually the best to photograph, as it is meant to be viewed from the side.

2. Insert your flowers, rotating the vase in quarter turns after each new insertion and making sure to fill the outside edge (see images in Steps 5 and 6).

3. Make sure you choose an appropriately attention-getting vessel, like the tall, V-shaped gold option we've chosen here. Contemporary and avant-garde arrangements are meant to stand out, and lower vases tend to be more basic.

STEP 1: By now, you know that your first step with any arrangement is to place your hands on each side of your vase to get a sense of how large your arrangement should be. In this case, I've added a few stems to further help myself visualize how wide the final design should be.

STEP 2: Start adding a quarter of your flowers around the outside edge of the vase, and then follow with a quarter of your greenery.

STEP 3: In this case, I'm using two types of greenery, so I'm adding each item one at a time.

STEP 4: Now that we've got some of our greens in place, it's time to go back to our flowers. I'm inserting each flower one at a time, turning the vase slightly before inserting another flower. Repeat this process until the edge of the vase is completely filled.

STEP 5: Finally, we'll use the remainder of our foliage to fill out the top of the arrangement, making sure to address any empty spaces.

STEP 6: One more chance for last looks and final touches. Always be sure to take one final look at your arrangement. The final stages of creating a flower arrangement are usually the best times to add a special element or elements that are too delicate to use in the beginning. In this case, we are adding some grasses.

Classic Contemporary Style

What comes to mind when you think of classic flower arrangements? When doing the research for this book, I tried to determine the industry standard for classic style and yielded confusing results. It seems that everyone has a different idea of what "classic" means. I've found that in order to be considered classic in style, a flower arrangement needs to focus on the following rough set of design principles: proportion, scale, harmony, rhythm, unity, and emphasis.

1. As always, start with your vase selection. No matter what material elements you select, they should be roughly the same size as your vase, creating a 1:1 ratio between the flowers and foliage. In this case, we are using a 5 × 5 inch cylinder vase. It's a classic.

2. Classic flower arrangements lend themselves to timeless blooms as well as popular varieties and cut flowers that continue to be in use today; think roses, hydrangeas, salal, lilies, carnations, and any other type of flower with a flat profile.

3. When making this arrangement, be sure to follow the basic principle of working from the outside edge to the middle. In this case we've chosen to work with roses and carnations.

STEP 1: For our classic arrangement we've chosen to one color to work with: green. As you can see here, we're gathering all of our materials in one hand as a way to gauge how large our arrangement will be. The amount of material you can hold in one hand will generally correspond to the size of your final arrangement.

STEP 2: Now it's time to start inserting your flowers. As always, begin by adding about a quarter of your flowers, working in a clockwise motion, and making sure that they are evenly distributed.

STEP 3: Next, begin adding your supplementary materials. Again, work in a clockwise motion, turning your vase after each insertion. Make sure you're paying close attention to your proportions as you go. Are all of your materials dispersed equally around your vase?

STEP 4: Now go in and add your final materials to the top of your arrangement, making sure that the center and the top of your design are just as full as the sides of your vase.

STEP 5: At this point, you should have a lovely classic arrangement. If you want to go further, you could add in some complementary colors. Here, we've decided to add in a few white variations to create a depth of color and texture.

STEP 6: Working in a circular motion, I'm dotting the top and the sides of my arrangement with more white flowers, making sure to place them evenly throughout. Now you have a gorgeous classic arrangement incorporating a few complementary hints of white as shown here.

Wild Organic Style

This look uses many different varieties of flowers, ranging from five different types of foliage and 12 to 16 different flowers, which will yield an incredible outcome.

1. As the name suggests, this style of flower arrangement offers an ideal moment to let your creativity take control . . . within reason. This is a chance to just let everything go and use whatever you have on hand. Once you've got your materials picked out, you can create any shape you choose!

2. Now is also an ideal time to use a less conventional vase. We've chosen a flat, rectangular vase for this arrangement, but you could use any number of shapes and varieties as long as they offer you enough space to create a voluminous arrangement. When using an unconventional container, make sure to test its ability to hold water first. There's nothing worse than finishing an arrangement only to find out that your vessel leaks.

3. The number one rule with these arrangements is excess! It's right there in the name: "wild." You want to give the viewer a sense of opulence and exuberance. So pile on the flowers and the foliage. Just do your best not to be too concerned with form. Instead, let the materials do all the talking.

STEP 1: For this arrangement we've chosen a relatively short vessel that will need some help to hold all of the different materials we're going to fill it with. As a result, we'll need to utilize some waterproof tape and some chicken wire. This will help everything stand up and stay in place.

STEP 2: Inserting cuttings from your houseplants is always fun. Here, we are using house palms and calathea leaves.

STEP 3: The principle here is simple. Fill the bottom of the vase with a rectangular mass of chicken wire, and tape the wire into place. Once your flowers are in place, no one will be able to see the tape or the wire lurking underneath.

STEP 4: Now it's time to start building our arrangement. I've opted for a mix of green, white, and cream for this exercise. I'm beginning by building a structure out of multiple types of foliage that will ring the outside of my rectangular vase. This will help to hold up the more delicate flowers that I'm going to add next.

STEP 5: With my basic framework in place, I now begin to add some of my more delicate pieces, making sure to place them around the edges and in places where they can be seen. This will create a striking balance between the greenery of my foliage and the delicate and crisp white blooms I've chosen to set against them.

Building on the Basics

Now that you've got a general sense of what the basic styles of flower arranging are, I'm going to throw you a little twist! These styles can be very helpful by providing you with a framework to follow when making an arrangement, helping to organize your thoughts and materials as you work toward a specific goal. However, consider these styles to serve only as a frame of reference. Mixing two different styles can yield spectacular results, so don't feel too confined!

Note: The following steps and Pro Tips apply when necessary to *all* arrangements.

STEP 1: As you begin, make sure not to cut the stems too short. Leave everything too long and flopping around, especially for garden-style arrangements.

STEP 2: Once your flowers are cut, insert a quarter of your total flowers and foliage into your vase, turning the vase as you add each stem. Don't bunch all of your stems up in one area or you will end up with an arrangement that looks messy and unintentional. Unintentional nature looks wrong, or off.

STEP 3: Be sure to keep your star flowers—that is, any flowers you want to display prominently and any flowers that are particularly delicate—for the end. Once you've placed the first quarter of your flowers, step back for a look. Ask yourself: Are the stems too short or too long? At this point, the answer should be too long.

STEP 4: Add another quarter of your materials, and step back for another look. Again, ask yourself if the stems are too short or too long.

STEP 5: Finally, add the last half of your materials. This will include all of your most delicate flowers and any flowers you want to display prominently. At this point, you should trim any stems that are too long.

Once you've got a handle on the basics of flower arranging, you can create a simple, elegant arrangement or make something wild and garden inspired. The possibilities are endless!

PRO TIP: When inserting flowers into the vase, be sure to only hold their stems and not push the flowers in. Sometimes, even if the flower looks like it is unharmed by touching, several hours later it will appear crushed.

PRO TIP: Keep several ferns and houseplants on hand that are good to clip from. Often arrangements just want that little something extra, and you won't have to look far if you have some options nearby.

Conclusion

To be a great florist can mean a great many things, but loving flowers is for everyone. There are not right or wrong, perfect or flawed, successful or unsuccessful flower arrangements. If you like the way an arrangement looks and it brings you joy, then it is perfect. This book is filled with tips to help at-home flower arrangers get more out of their work, and to demystify some of the more mysterious aspects of the flower industry. In addition, this book is also meant as a kind of philosophical guide to the world of flowers. As long as you approach an arrangement with joy, there's no way you'll end up with a finished product that you're unhappy with.

When my editors asked me to put this book together, I was first filled with a sense of excitement. I was excited by the opportunity to share some of the FlowerSchool philosophy with readers who have not yet made the trip to New York or Los Angeles to see us. However, after I put the first few words down on paper, I immediately began to feel a less pleasant sensation: I was overwhelmed. There are so many details that I have mastered over the years, from picking colors to picking flowers and understanding the way each flower enters a vase. How would I be able to explain every detail? The answer is: I couldn't. Flower arranging is about accepting the fact that Mother Nature has her own ideas. Nature is filled with variables, and I only had a couple hundred pages in which to confront them. So instead of trying to create a kind of cookbook for flower arrangers, with recipes and precise measurements that just wouldn't make sense in the real world, I've opted for a more philosophical look at doing flowers.

You will never find the same flower twice, so rather than focusing on the flower of your dreams, focus instead on approaching each arrangement with an open mind. And then see what comes to you. If you start to work on a mixed arrangement of peonies and lilacs and suddenly you find that the peonies don't look great, well, pull the lilacs forward and bury the peonies. If the lilacs don't look that great and the peonies are spectacular, then put the peonies on top. Then, once you've made your decision, move on and let the flowers do all the talking.

Doing flowers is a naturally poetic form of personal expression. The colors you choose, the style you adopt, and the vase you select can all affect the outcome of your arrangement. Making these decisions can often seem like a Herculean task. But I urge you not to overthink things. Making a flower arrangement is like making a salad. If you want walnuts but find there are none available, then use some other nut. You may discover a newfound love of cashews! If you learn just one thing from this book, I hope it's this: Enjoy what you do and experiment often.

Acknowledgments

There have been so many wonderful people whom I have worked with over the years.

Thank you to Eileen Johnson for letting us retool this school into the institution that it is today. It will hopefully live on forever. Brittany Lenig, my co-conspirator, thank you for always supporting the FlowerSchool mission no matter how bizarre it seemed at the time. We have done so many wonderful things and there are still so many more dreams that have yet to come true.

I would like to thank the wonderful FlowerSchool Master Florists and designers whom I have had the honor to work with over the years. Michael George, Remco Van Vliet, Ariella Chezar, Takaya Sato, Lewis Miller, Christian Tortu, Shane Connolly, Meredith Perez, Kiana Underwood, Oscar Mora, Olivier Giugni, and Ingrid Carozzi. Thank you for showing us what the best of the best can do. Being a florist is no easy job, and it's been a privilege to help share your vision with the FlowerSchool community.

Thank you to all the passionate florists who have contributed to FlowerSchool's Floral Design Program, including Meghan Riley, Beth Horta, Juan Villanueva, Laura Seita, Barbara Mele, Jessi Owens, Tom Sebenius, and Jin Park. Without your enthusiasm for our mission, we would not be in the position we are in now. Thank you.

At the foundation of this industry are the growers, distributors, and wholesalers. All the players in the 28th Street wholesale district, like Paul from J&P; Gary and staff of G.Page Wholesale; Cas, Peet, Ed, and Chris of Dutch Flower Line; Gus and Ed at US Evergreen; Bas and Marco at Hilverda da Boer; Joe and Michael at FleuraMetz; and Dave and Daryl at Rallis. Six days a week we are able to get flower deliveries for any and every project we are working on. Without their dedication to the field, nothing would be possible.

Finally, a big thank you to Accent Decor for supporting FlowerSchool for so many years. Providing access to so many wonderful containers for our students to try out has given them so much confidence.

Index

Photos are in **bold**.

About the Author

CALVERT CRARY is the executive director of FlowerSchool NY and FlowerSchool LA. Formerly a fashion and editorial photographer, Calvert made the transition to florist and then to floral entrepreneur, having opened and successfully run three flower businesses in New York City. Calvert has trained and coached many students to open shops and reorganize existing floral businesses into thriving careers.